A concise introduction to the theory of numbers

A concise introduction to the theory of numbers

ALAN BAKER
Professor of Pure Mathematics in the University of Cambridge

PUBLISHED BY THE PRESS SYNDICATE OF THE UNIVERSITY OF CAMBRIDGE
The Pitt Building, Trumpington Street, Cambridge CB2 1RP, United Kingdom

CAMBRIDGE UNIVERSITY PRESS
The Edinburgh Building, Cambridge CB2 2RU, United Kingdom
40 West 20th Street, New York, NY 10011-4211, USA
10 Stamford Road, Oakleigh, Melbourne 3166, Australia

First published 1984
Reprinted 1986 (twice), 1988, 1990, 1994, 1997

A catalogue record for this book is available from the British Library

Library of Congress catalogue card number: 84–1911

ISBN 0 521 28654 9 paperback

Transferred to digital printing 2002

Contents

Preface ix

Introduction: Gauss and number theory xi

1 Divisibility 1
1 Foundations 1
2 Division algorithm 1
3 Greatest common divisor 2
4 Euclid's algorithm 3
5 Fundamental theorem 3
6 Properties of the primes 4
7 Further reading 6
8 Exercises 7

2 Arithmetical functions 8
1 The function $[x]$ 8
2 Multiplicative functions 9
3 Euler's (totient) function $\phi(n)$ 9
4 The Möbius function $\mu(n)$ 10
5 The functions $\tau(n)$ and $\sigma(n)$ 11
6 Average orders 12
7 Perfect numbers 14
8 The Riemann zeta-function 14
9 Further reading 16
10 Exercises 16

3 Congruences 18
1 Definitions 18
2 Chinese remainder theorem 18
3 The theorems of Fermat and Euler 19
4 Wilson's theorem 20

5 Lagrange's theorem 21
6 Primitive roots 22
7 Indices 24
8 Further reading 25
9 Exercises 25

4 Quadratic residues 27
1 Legendre's symbol 27
2 Euler's criterion 27
3 Gauss' lemma 28
4 Law of quadratic reciprocity 29
5 Jacobi's symbol 31
6 Further reading 32
7 Exercises 33

5 Quadratic forms 35
1 Equivalence 35
2 Reduction 36
3 Representations by binary forms 37
4 Sums of two squares 38
5 Sums of four squares 39
6 Further reading 41
7 Exercises 41

6 Diophantine approximation 43
1 Dirichlet's theorem 43
2 Continued fractions 44
3 Rational approximations 46
4 Quadratic irrationals 48
5 Liouville's theorem 50
6 Transcendental numbers 53
7 Minkowski's theorem 56
8 Further reading 59
9 Exercises 59

7 Quadratic fields 61
1 Algebraic number fields 61
2 The quadratic field 62
3 Units 63
4 Primes and factorization 65
5 Euclidean fields 67

6 The Gaussian field 69
7 Further reading 71
8 Exercises 72

8 Diophantine equations 74
1 The Pell equation 74
2 The Thue equation 77
3 The Mordell equation 79
4 The Fermat equation 84
5 The Catalan equation 87
6 Further reading 89
7 Exercises 90

Preface

It has been customary in Cambridge for many years to include as part of the Mathematical Tripos a brief introductory course on the Theory of Numbers. This volume is a somewhat fuller version of the lecture notes attaching to the course as delivered by me in recent times. It has been prepared on the suggestion and with the encouragement of the University Press.

The subject has a long and distinguished history, and indeed the concepts and problems relating to the theory have been instrumental in the foundation of a large part of mathematics. The present text describes the rudiments of the field in a simple and direct manner. It is very much to be hoped that it will serve to stimulate the reader to delve into the rich literature associated with the subject and thereby to discover some of the deep and beautiful theories that have been created as a result of numerous researches over the centuries. Some guides to further study are given at the ends of the chapters. By way of introduction, there is a short account of the *Disquisitiones arithmeticae* of Gauss, and, to begin with, the reader can scarcely do better than to consult this famous work.

I am grateful to Mrs S. Lowe for her careful preparation of the typescript, to Mr P. Jackson for his meticulous subediting, to Dr D. J. Jackson for providing me with a computerized version of Fig. 8.1, and to Dr R. C. Mason for his help in checking the proof-sheets and for useful suggestions.

Cambridge 1983 A.B.

Introduction

*Gauss and number theory**

Without doubt the theory of numbers was Gauss' favourite subject. Indeed, in a much quoted dictum, he asserted that Mathematics is the Queen of the Sciences and the Theory of Numbers is the Queen of Mathematics. Moreover, in the introduction to Eisenstein's *Mathematische Abhandlungen*, Gauss wrote 'The Higher Arithmetic presents us with an inexhaustible storehouse of interesting truths – of truths, too, which are not isolated but stand in the closest relation to one another, and between which, with each successive advance of the science, we continually discover new and sometimes wholly unexpected points of contact. A great part of the theories of Arithmetic derive an additional charm from the peculiarity that we easily arrive by induction at important propositions which have the stamp of simplicity upon them but the demonstration of which lies so deep as not to be discovered until after many fruitless efforts; and even then it is obtained by some tedious and artificial process while the simpler methods of proof long remain hidden from us.'

All this is well illustrated by what is perhaps Gauss' most profound publication, namely his *Disquisitiones arithmeticae*. It has been described, quite justifiably I believe, as the Magna Carta of Number Theory, and the depth and originality of thought manifest in this work are particularly remarkable considering that it was written when Gauss was only about eighteen years of age. Of course, as Gauss said himself, not all of the subject matter was new at the time of writing, and Gauss

* This article was originally prepared for a meeting of the British Society for the History of Mathematics held in Cambridge in 1977 to celebrate the bicentenary of Gauss' birth.

acknowledged the considerable debt that he owed to earlier scholars, in particular Fermat, Euler, Lagrange and Legendre. But the *Disquisitiones arithmeticae* was the first systematic treatise on the Higher Arithmetic and it provided the foundations and stimulus for a great volume of subsequent research which is in fact continuing to this day. The importance of the work was recognized as soon as it was published in 1801 and the first edition quickly became unobtainable; indeed many scholars of the time had to resort to taking handwritten copies. But it was generally regarded as a rather impenetrable work and it was probably not widely understood; perhaps the formal latin style contributed in this respect. Now, however, after numerous reformulations, most of the material is very well known, and the earlier sections at least are included in every basic course on number theory.

The text begins with the definition of a congruence, namely two numbers are said to be congruent modulo n if their difference is divisible by n. This is plainly an equivalence relation in the now familiar terminology. Gauss proceeds to the discussion of linear congruences and shows that they can in fact be treated somewhat analogously to linear equations. He then turns his attention to power residues and introduces, amongst other things, the concepts of primitive roots and indices; and he notes, in particular, the resemblance between the latter and the ordinary logarithms. There follows an exposition of the theory of quadratic congruences, and it is here that we meet, more especially, the famous law of quadratic reciprocity; this asserts that if p, q are primes, not both congruent to 3 (mod 4), then p is a residue or non-residue of q according as q is a residue or non-residue of p, while in the remaining case the opposite occurs. As is well known, Gauss spent a great deal of time on this result and gave several demonstrations; and it has subsequently stimulated much excellent research. In particular, following works of Jacobi, Eisenstein and Kummer, Hilbert raised as the ninth of his famous list of problems presented at the Paris Congress of 1900 the question of obtaining higher reciprocity laws, and this led to the celebrated studies of Furtwängler, Artin and others in the context of class field theory.

By far the largest section of the *Disquisitiones arithmeticae* is concerned with the theory of binary quadratic forms. Here Gauss describes how quadratic forms with a given discriminant can be divided into classes so that two forms belong to the same class if and only if there exists an integral unimodular substitution relating them, and how the classes can be divided into genera, so that two forms are in the same genus if and only if they are rationally equivalent. He proceeds to apply these concepts so as, for instance, to throw light on the difficult question of the representation of integers by binary forms. It is a remarkable and beautiful theory with many important ramifications. Indeed, after re-interpretation in terms of quadratic fields, it became apparent that it could be applied much more widely, and in fact it can be regarded as having provided the foundations for the whole of algebraic number theory. The term Gaussian field, meaning the field generated over the rationals by i, is a reminder of Gauss' pioneering work in this area.

The remainder of the *Disquisitiones arithmeticae* contains results of a more miscellaneous character, relating, for instance, to the construction of seventeen-sided polygons, which was clearly of particular appeal to Gauss, and to what is now termed the cyclotomic field, that is the field generated by a primitive root of unity. And especially noteworthy here is the discussion of certain sums involving roots of unity, now referred to as Gaussian sums, which play a fundamental role in the analytic theory of numbers.

I conclude this introduction with some words of Mordell. In an essay published in 1917 he wrote 'The theory of numbers is unrivalled for the number and variety of its results and for the beauty and wealth of its demonstrations. The Higher Arithmetic seems to include most of the romance of mathematics. As Gauss wrote to Sophie Germain, the enchanting beauties of this sublime study are revealed in their full charm only to those who have the courage to pursue it.' And Mordell added 'We are reminded of the folk-tales, current amongst all peoples, of the Prince Charming who can assume his proper form as a handsome prince only because of the devotedness of the faithful heroine.'

1

Divisibility

1 Foundations

The set 1, 2, 3, . . . of all natural numbers will be denoted by $\mathbb{N}$. There is no need to enter here into philosophical questions concerning the existence of $\mathbb{N}$. It will suffice to assume that it is a given set for which the Peano axioms are satisfied. They imply that addition and multiplication can be defined on $\mathbb{N}$ such that the commutative, associative and distributive laws are valid. Further, an ordering on $\mathbb{N}$ can be introduced so that either $m<n$ or $n<m$ for any distinct elements m, n in $\mathbb{N}$. Furthermore, it is evident from the axioms that the principle of mathematical induction holds and that every non-empty subset of $\mathbb{N}$ has a least member. We shall frequently appeal to these properties.

As customary, we shall denote by $\mathbb{Z}$ the set of integers $0, \pm 1, \pm 2, \ldots$, and by $\mathbb{Q}$ the set of rationals, that is the numbers p/q with p in $\mathbb{Z}$ and q in $\mathbb{N}$. The construction, commencing with $\mathbb{N}$, of $\mathbb{Z}$, $\mathbb{Q}$ and then the real and complex numbers $\mathbb{R}$ and $\mathbb{C}$ forms the basis of Mathematical Analysis and it is assumed known.

2 Division algorithm

Suppose that a, b are elements of $\mathbb{N}$. One says that b divides a (written $b|a$) if there exists an element c of $\mathbb{N}$ such that $a=bc$. In this case b is referred to as a divisor of a, and a is called a multiple of b. The relation $b|a$ is reflexive and transitive but not symmetric; in fact if $b|a$ and $a|b$ then $a=b$. Clearly also if $b|a$ then $b\leq a$ and so a natural number has only finitely many divisors. The concept of divisibility is readily extended to $\mathbb{Z}$; if a, b are elements of $\mathbb{Z}$, with $b \neq 0$, then b is said to divide a if there exists c in $\mathbb{Z}$ such that $a = bc$.

We shall frequently appeal to the division algorithm. This asserts that for any a, b in $\mathbb{Z}$, with $b > 0$, there exist q, r in $\mathbb{Z}$ such that $a = bq + r$ and $0 \leq r < b$. The proof is simple; indeed if bq is the largest multiple of b that does not exceed a then the integer $r = a - bq$ is certainly non-negative and, since $b(q+1) > a$, we have $r < b$. The result plainly remains valid for any integer $b \neq 0$ provided that the bound $r < b$ is replaced by $r < |b|$.

3 Greatest common divisor

By the greatest common divisor of natural numbers a, b we mean an element d of $\mathbb{N}$ such that $d|a$, $d|b$ and every common divisor of a and b also divides d. We proceed to prove that a number d with these properties exists; plainly it will be unique, for any other such number d' would divide a, b and so also d, and since similarly $d|d'$ we have $d = d'$.

Accordingly consider the set of all natural numbers of the form $ax + by$ with x, y in $\mathbb{Z}$. The set is not empty since, for instance, it contains a and b; hence there is a least member d, say. Now $d = ax + by$ for some integers x, y, whence every common divisor of a and b certainly divides d. Further, by the division algorithm, we have $a = dq + r$ for some q, r in $\mathbb{Z}$ with $0 \leq r < d$; this gives $r = ax' + by'$, where $x' = 1 - qx$ and $y' = -qy$. Thus, from the minimal property of d, it follows that $r = 0$ whence $d|a$. Similarly we have $d|b$, as required.

It is customary to signify the greatest common divisor of a, b by (a, b). Clearly, for any n in $\mathbb{N}$, the equation $ax + by = n$ is soluble in integers x, y if and only if (a, b) divides n. In the case $(a, b) = 1$ we say that a and b are relatively prime or coprime (or that a is prime to b). Then the equation $ax + by = n$ is always soluble.

Obviously one can extend these concepts to more than two numbers. In fact one can show that any elements $a_1, \ldots, a_m$ of $\mathbb{N}$ have a greatest common divisor $d = (a_1, \ldots, a_m)$ such that $d = a_1x_1 + \cdots + a_mx_m$ for some integers $x_1, \ldots, x_m$. Further, if $d = 1$, we say that $a_1, \ldots, a_m$ are relatively prime and then the equation $a_1x_1 + \cdots + a_mx_m = n$ is always soluble.

4 Euclid's algorithm

A method for finding the greatest common divisor d of a, b was described by Euclid. It proceeds as follows.

By the division algorithm there exist integers q_1, r_1 such that $a = bq_1 + r_1$ and $0 \leq r_1 < b$. If $r_1 \neq 0$ then there exist integers q_2, r_2 such that $b = r_1 q_2 + r_2$ and $0 \leq r_2 < r_1$. If $r_2 \neq 0$ then there exist integers q_3, r_3 such that $r_1 = r_2 q_3 + r_3$ and $0 \leq r_3 < r_2$. Continuing thus, one obtains a decreasing sequence $r_1, r_2, \ldots$ satisfying $r_{j-2} = r_{j-1} q_j + r_j$. The sequence terminates when $r_{k+1} = 0$ for some k, that is when $r_{k-1} = r_k q_{k+1}$. It is then readily verified that $d = r_k$. Indeed it is evident from the equations that every common divisor of a and b divides $r_1, r_2, \ldots, r_k$; and moreover, viewing the equations in the reverse order, it is clear that r_k divides each r_j and so also b and a.

Euclid's algorithm furnishes another proof of the existence of integers x, y satisfying $d = ax + by$, and furthermore it enables these x, y to be explicitly calculated. For we have $d = r_k$ and $r_j = r_{j-2} - r_{j-1} q_j$ whence the required values can be obtained by successive substitution. Let us take, for example, $a = 187$ and $b = 35$. Then, following Euclid, we have

$$187 = 35 \cdot 5 + 12, \quad 35 = 12 \cdot 2 + 11, \quad 12 = 11 \cdot 1 + 1.$$

Thus we see that $(187, 35) = 1$ and moreover

$$1 = 12 - 11 \cdot 1 = 12 - (35 - 12 \cdot 2) = 3(187 - 35 \cdot 5) - 35.$$

Hence a solution of the equation $187x + 35y = 1$ in integers x, y is given by $x = 3$, $y = -16$.

There is a close connection between Euclid's algorithm and the theory of continued fractions; this will be discussed in Chapter 6.

5 Fundamental theorem

A natural number, other than 1, is called a prime if it is divisible only by itself and 1. The smallest primes are therefore given by 2, 3, 5, 7, 11, ...

Let n be any natural number other than 1. The least divisor of n that exceeds 1 is plainly a prime, say p_1. If $n \neq p_1$ then, similarly, there is a prime p_2 dividing n/p_1. If $n \neq p_1 p_2$ then there is a prime p_3 dividing $n/p_1 p_2$; and so on. After a finite number of steps we obtain $n = p_1 \cdots p_m$; and by grouping together we get the standard factorization (or canonical decomposition) $n = p_1^{j_1} \cdots p_k^{j_k}$, where $p_1, \ldots, p_k$ denote distinct primes and $j_1, \ldots, j_k$ are elements of $\mathbb{N}$.

The fundamental theorem of arithmetic asserts that the above factorization is unique except for the order of the factors. To prove the result, note first that if a prime p divides a product mn of natural numbers then either p divides m or p divides n. Indeed if p does not divide m then $(p, m) = 1$ whence there exist integers x, y such that $px + my = 1$; thus we have $pnx + mny = n$ and hence p divides n. More generally we conclude that if p divides $n_1 n_2 \cdots n_k$ then p divides n_l for some l. Now suppose that, apart from the factorization $n = p_1^{j_1} \cdots p_k^{j_k}$ derived above, there is another decomposition and that p' is one of the primes occurring therein. From the preceding conclusion we obtain $p' = p_l$ for some l. Hence we deduce that, if the standard factorization for n/p' is unique, then so also is that for n. The fundamental theorem follows by induction.

It is simple to express the greatest common divisor (a, b) of elements a, b of $\mathbb{N}$ in terms of the primes occurring in their decompositions. In fact we can write $a = p_1^{\alpha_1} \cdots p_k^{\alpha_k}$ and $b = p_1^{\beta_1} \cdots p_k^{\beta_k}$, where $p_1, \ldots, p_k$ are distinct primes and the αs and βs are non-negative integers; then $(a, b) = p_1^{\gamma_1} \cdots p_k^{\gamma_k}$, where $\gamma_l = \min(\alpha_l, \beta_l)$. With the same notation, the lowest common multiple of a, b is defined by $\{a, b\} = p_1^{\delta_1} \cdots p_k^{\delta_k}$, where $\delta_l = \max(\alpha_l, \beta_l)$. The identity $(a, b)\{a, b\} = ab$ is readily verified.

6 Properties of the primes

There exist infinitely many primes, for if $p_1, \ldots, p_n$ is any finite set of primes then $p_1 \cdots p_n + 1$ is divisible by a prime different from $p_1, \ldots, p_n$; the argument is due to Euclid. It follows that, if p_n is the nth prime in ascending order of magnitude, then p_m divides $p_1 \cdots p_n + 1$ for some $m \geq n+1$; from this we deduce by induction that $p_n < 2^{2^n}$. In fact a much stronger result is known; indeed $p_n \sim n \log n$ as $n \to \infty$.† The result is equivalent to the assertion that the number $\pi(x)$ of primes $p \leq x$ satisfies $\pi(x) \sim x/\log x$ as $x \to \infty$. This is called the prime-number

† The notation $f \sim g$ means that $f/g \to 1$; and one says that f is asymptotic to g.

theorem and it was proved by Hadamard and de la Vallée Poussin independently in 1896. Their proofs were based on properties of the Riemann zeta-function about which we shall speak in Chapter 2. In 1737 Euler proved that the series $\sum 1/p_n$ diverges and he noted that this gives another demonstration of the existence of infinitely many primes. In fact it can be shown by elementary arguments that, for some number c,

$$\sum_{p\le x} 1/p = \log\log x + c + O(1/\log x).$$

Fermat conjectured that the numbers $2^{2^n}+1$ ($n=1, 2, \ldots$) are all primes; this is true for $n=1, 2, 3$ and 4 but false for $n=5$, as was proved by Euler. In fact 641 divides $2^{32}+1$. Numbers of the above form that are primes are called Fermat primes. They are closely connected with the existence of a construction of a regular plane polygon with ruler and compasses only. In fact the regular plane polygon with p sides, where p is a prime, is capable of construction if and only if p is a Fermat prime. It is not known at present whether the number of Fermat primes is finite or infinite.

Numbers of the form 2^n-1 that are primes are called Mersenne primes. In this case n is a prime, for plainly 2^m-1 divides 2^n-1 if m divides n. Mersenne primes are of particular interest in providing examples of large prime numbers; for instance it is known that $2^{44\,497}-1$ is the 27th Mersenne prime, a number with 13 395 digits.

It is easily seen that no polynomial $f(n)$ with integer coefficients can be prime for all n in $\mathbb{N}$, or even for all sufficiently large n, unless f is constant. Indeed by Taylor's theorem, $f(mf(n)+n)$ is divisible by $f(n)$ for all m in $\mathbb{N}$. On the other hand, the remarkable polynomial n^2-n+41 is prime for $n=1, 2, \ldots, 40$. Furthermore one can write down a polynomial $f(n_1, \ldots, n_k)$ with the property that, as the n_j run through the elements of $\mathbb{N}$, the set of positive values assumed by f is precisely the sequence of primes. The latter result arises from studies in logic relating to Hilbert's tenth problem (see Chapter 8).

The primes are well distributed in the sense that, for every $n>1$, there is always a prime between n and $2n$. This result, which is commonly referred to as Bertrand's postulate, can be regarded as the forerunner of extensive researches on the difference $p_{n+1}-p_n$ of consecutive primes. In fact estimates of the form $p_{n+1}-p_n=O(p_n^{\kappa})$ are known with values of κ just a little greater than $\frac{1}{2}$; but, on the other hand, the difference is certainly not bounded, since the consecutive integers $n!+m$ with $m=2, 3, \ldots, n$ are all composite. A famous theorem of Dirichlet asserts that any arithmetical progression a, $a+q$, $a+2q, \ldots$, where $(a, q)=1$, contains infinitely many primes. Some special cases, for instance the existence of infinitely many primes of the form $4n+3$, can be deduced simply by modifying Euclid's argument given at the beginning, but the general result lies quite deep. Indeed Dirichlet's proof involved, amongst other things, the concepts of characters and L-functions, and of class numbers of quadratic forms, and it has been of far-reaching significance in the history of mathematics.

Two notorious unsolved problems in prime-number theory are the Goldbach conjecture, mentioned in a letter to Euler of 1742, to the effect that every even integer (>2) is the sum of two primes, and the twin-prime conjecture, to the effect that there exist infinitely many pairs of primes, such as 3, 5 and 17, 19, that differ by 2. By ingenious work on sieve methods, Chen showed in 1974 that these conjectures are valid if one of the primes is replaced by a number with at most two prime factors (assuming, in the Goldbach case, that the even integer is sufficiently large). The oldest known sieve, incidentally, is due to Eratosthenes. He observed that if one deletes from the set of integers $2, 3, \ldots, n$, first all multiples of 2, then all multiples of 3, and so on up to the largest integer not exceeding $\sqrt{n}$, then only primes remain. Studies on Goldbach's conjecture gave rise to the Hardy–Littlewood circle method of analysis and, in particular, to the celebrated theorem of Vinogradov to the effect that every sufficiently large odd integer is the sum of three primes.

7 Further reading

For a good account of the Peano axioms see E. Landau, *Foundations of analysis* (Chelsea Publ. Co., New York, 1951).

The division algorithm, Euclid's algorithm and the fundamental theorem of arithmetic are discussed in every elementary text on number theory. The tracts are too numerous to list here but for many years the book by G. H. Hardy and E. M. Wright, *An introduction to the theory of numbers* (Oxford U.P., 5th edn, 1979) has been regarded as a standard work in the field. The books of similar title by T. Nagell (Wiley, New York, 1951) and H. M. Stark (MIT Press, Cambridge, Mass., 1978) are also to be recommended, as well as the volume by E. Landau, *Elementary number theory* (Chelsea Publ. Co., New York, 1958).

For properties of the primes, see the book by Hardy and Wright mentioned above and, for more advanced reading, see, for instance, H. Davenport, *Multiplicative number theory* (Springer-Verlag, Berlin, 2nd ed, 1980) and H. Halberstam and H. E. Richert, *Sieve methods* (Academic Press, London and New York, 1974). The latter contains, in particular, a proof of Chen's theorem. The result referred to on a polynomial in several variables representing primes arose from work of Davis, Robinson, Putnam and Matiyasevich on Hilbert's tenth problem; see, for instance, the article in *American Math. Monthly* **83** (1976), 449–64, where it is shown that 12 variables suffice.

8 Exercises

(i) Find integers x, y such that $95x+432y=1$.

(ii) Find integers x, y, z such that $35x+55y+77z=1$.

(iii) Prove that $1+\frac{1}{2}+\cdots+1/n$ is not an integer for $n>1$.

(iv) Prove that
$$(\{a, b\}, \{b, c\}, \{c, a\}) = \{(a, b), (b, c), (c, a)\}.$$

(v) Prove that if $g_1, g_2, \ldots$ are integers >1 then every natural number can be expressed uniquely in the form $a_0+a_1g_1+a_2g_1g_2+\cdots+a_kg_1\cdots g_k$, where the a_j are integers satisfying $0\leq a_j<g_{j+1}$.

(vi) Show that there exist infinitely many primes of the form $4n+3$.

(vii) Show that, if 2^n+1 is a prime then it is in fact a Fermat prime.

(viii) Show that, if $m>n$, then $2^{2^n}+1$ divides $2^{2^m}-1$ and so $(2^{2^m}+1, 2^{2^n}+1)=1$.

(ix) Deduce that $p_{n+1}\leq 2^{2^n}+1$, whence $\pi(x)\geq \log\log x$ for $x\geq 2$.

2

Arithmetical functions

1 The function $[x]$

For any real x, one signifies by $[x]$ the largest integer $\leq x$, that is, the unique integer such that $x-1<[x]\leq x$. The function is called 'the integral part of x'. It is readily verified that $[x+y]\geq[x]+[y]$ and that, for any positive integer n, $[x+n]=[x]+n$ and $[x/n]=[[x]/n]$. The difference $x-[x]$ is called 'the fractional part of x'; it is written $\{x\}$ and satisfies $0\leq\{x\}<1$.

Let now p be a prime. The largest integer l such that p^l divides $n!$ can be neatly expressed in terms of the above function. In fact, on noting that $[n/p]$ of the numbers $1, 2, \ldots, n$ are divisible by p, that $[n/p^2]$ are divisible by p^2, and so on, we obtain

$$l=\sum_{m=1}^{n}\sum_{\substack{j=1\\ p^j|m}}^{\infty}1=\sum_{j=1}^{\infty}\sum_{\substack{m=1\\ p^j|m}}^{n}1=\sum_{j=1}^{\infty}[n/p^j].$$

It follows easily that $l\leq[n/(p-1)]$; for the latter sum is at most $n(1/p+1/p^2+\cdots)$. The result also shows at once that the binomial coefficient

$$\binom{m}{n}=\frac{m!}{n!\,(m-n)!}$$

is an integer; for we have

$$[m/p^j]\geq[n/p^j]+[(m-n)/p^j].$$

Indeed, more generally, if $n_1, \ldots, n_k$ are positive integers such that $n_1+\cdots+n_k=m$ then the expression $m!/(n_1!\cdots n_k!)$ is an integer.

2 Multiplicative functions

A real function f defined on the positive integers is said to be multiplicative if $f(m)f(n)=f(mn)$ for all m, n with $(m, n)=1$. We shall meet many examples. Plainly if f is multiplicative and does not vanish identically then $f(1)=1$. Further if $n=p_1^{j_1}\cdots p_k^{j_k}$ in standard form then

$$f(n)=f(p_1^{j_1})\cdots f(p_k^{j_k}).$$

Thus to evaluate f it suffices to calculate its values on the prime powers; we shall appeal to this property frequently.

We shall also use the fact that if f is multiplicative and if

$$g(n)=\sum_{d|n} f(d),$$

where the sum is over all divisors d of n, then g is a multiplicative function. Indeed, if $(m, n)=1$, we have

$$\begin{aligned} g(mn)&=\sum_{d|m}\sum_{d'|n} f(dd')=\sum_{d|m} f(d)\sum_{d'|n} f(d') \\ &=g(m)g(n). \end{aligned}$$

3 Euler's (totient) function $\phi(n)$

By $\phi(n)$ we mean the number of numbers $1, 2, \ldots, n$ that are relatively prime to n. Thus, in particular, $\phi(1)=\phi(2)=1$ and $\phi(3)=\phi(4)=2$.

We shall show, in the next chapter, from properties of congruences, that ϕ is multiplicative. Now, as is easily verified, $\phi(p^j)=p^j-p^{j-1}$ for all prime powers p^j. It follows at once that

$$\phi(n)=n\prod_{p|n}(1-1/p).$$

We proceed to establish this formula directly without assuming that ϕ is multiplicative. In fact the formula furnishes another proof of this property.

Let $p_1, \ldots, p_k$ be the distinct prime factors of n. Then it suffices to show that $\phi(n)$ is given by

$$n-\sum_r (n/p_r)+\sum_{r>s} n/(p_rp_s)-\sum_{r>s>t} n/(p_rp_sp_t)+\cdots.$$

But n/p_r is the number of numbers $1, 2, \ldots, n$ that are divisible by p_r, $n/(p_rp_s)$ is the number that are divisible by p_rp_s, and so on. Hence the above expression is

$$\sum_{m=1}^{n}\Big(1-\sum_{\substack{r\\ p_r|m}}1+\sum_{\substack{r>s\\ p_rp_s|m}}1-\cdots\Big)$$

$$=\sum_{m=1}^{n}\Big(1-\binom{l}{1}+\binom{l}{2}-\cdots\Big),$$

where $l=l(m)$ is the number of primes $p_1,\ldots,p_k$ that divide m. Now the summand on the right is $(1-1)^l=0$ if $l>0$, and it is 1 if $l=0$. The required result follows. The demonstration is a particular example of an argument due to Sylvester.

It is a simple consequence of the multiplicative property of ϕ that

$$\sum_{d|n}\phi(d)=n.$$

In fact the expression on the left is multiplicative and, when $n=p^j$, it becomes

$$\phi(1)+\phi(p)+\cdots+\phi(p^j)$$
$$=1+(p-1)+\cdots+(p^j-p^{j-1})=p^j.$$

4 The Möbius function $\mu(n)$

This is defined, for any positive integer n, as 0 if n contains a squared factor, and as $(-1)^k$ if $n=p_1\cdots p_k$ as a product of k distinct primes. Further, by convention, $\mu(1)=1$.

It is clear that μ is multiplicative. Thus the function

$$\nu(n)=\sum_{d|n}\mu(d)$$

is also multiplicative. Now for all prime powers p^j with $j>0$ we have $\nu(p^j)=\mu(1)+\mu(p)=0$. Hence we obtain the basic property, namely $\nu(n)=0$ for $n>1$ and $\nu(1)=1$. We proceed to use this property to establish the Möbius inversion formulae.

Let f be any arithmetical function, that is a function defined on the positive integers, and let

$$g(n)=\sum_{d|n}f(d).$$

Then we have

$$f(n)=\sum_{d|n}\mu(d)g(n/d).$$

In fact the right hand side is

$$\sum_{d|n}\sum_{d'|n/d}\mu(d)f(d')=\sum_{d'|n}f(d')\nu(n/d'),$$

and the result follows since $\nu(n/d')=0$ unless $d'=n$. The converse also holds, for we can write the second equation in the form

$$f(n)=\sum_{d'|n}\mu(n/d')g(d')$$

and then

$$\sum_{d|n}f(d)=\sum_{d|n}f(n/d)=\sum_{d|n}\sum_{d'|n/d}\mu(n/dd')g(d')$$
$$=\sum_{d'|n}g(d')\nu(n/d').$$

Again we have $\nu(n/d')=0$ unless $d'=n$, whence the expression on the right is $g(n)$.

The Euler and Möbius functions are related by the equation

$$\phi(n)=n\sum_{d|n}\mu(d)/d.$$

This can be seen directly from the formula for ϕ established in § 3, and it also follows at once by Möbius inversion from the property of ϕ recorded at the end of § 3. Indeed the relation is clear from the multiplicative properties of ϕ and μ.

There is an analogue of Möbius inversion for functions defined over the reals, namely if

$$g(x)=\sum_{n\le x}f(x/n)$$

then

$$f(x)=\sum_{n\le x}\mu(n)g(x/n).$$

In fact the last sum is

$$\sum_{n\le x}\sum_{m\le x/n}\mu(n)f(x/mn)=\sum_{l\le x}f(x/l)\nu(l)$$

and the result follows since $\nu(l)=0$ for $l>1$. We shall give several applications of Möbius inversion in the examples at the end of the chapter.

5 The functions $\tau(n)$ and $\sigma(n)$

For any positive integer n, we denote by $\tau(n)$ the number of divisors of n (in some books, in particular in that of Hardy and Wright, the function is written $d(n)$). By $\sigma(n)$ we denote the sum of the divisors of n. Thus

$$\tau(n)=\sum_{d|n} 1, \qquad \sigma(n)=\sum_{d|n} d.$$

It is plain that both $\tau(n)$ and $\sigma(n)$ are multiplicative. Further, for any prime power p^j we have $\tau(p^j)=j+1$ and

$$\sigma(p^j)=1+p+\cdots+p^j=(p^{j+1}-1)/(p-1).$$

Thus if p^j is the highest power of p that divides n then

$$\tau(n)=\prod_{p|n}(j+1), \qquad \sigma(n)=\prod_{p|n}(p^{j+1}-1)/(p-1).$$

It is easy to give rough estimates for the sizes of $\tau(n)$ and $\sigma(n)$. Indeed we have $\tau(n)<cn^{\delta}$ for any $\delta>0$, where c is a number depending only on δ; for the function $f(n)=\tau(n)/n^{\delta}$ is multiplicative and satisfies $f(p^j)=(j+1)/p^{j\delta}<1$ for all but a finite number of values of p and j, the exceptions being bounded in terms of δ. Further we have

$$\sigma(n)=n\sum_{d|n} 1/d\le n\sum_{d\le n} 1/d<n(1+\log n).$$

The last estimate implies that $\phi(n)>\frac{1}{4}n/\log n$ for $n>1$. In fact the function $f(n)=\sigma(n)\phi(n)/n^2$ is multiplicative and, for any prime power p^j, we have

$$f(p^j)=1-p^{-j-1}\ge 1-1/p^2;$$

hence, since

$$\prod_{p|n}(1-1/p^2)\ge\prod_{m=2}^{\infty}(1-1/m^2)=\tfrac{1}{2},$$

it follows that $\sigma(n)\phi(n)\ge\frac{1}{2}n^2$, and this together with $\sigma(n)<2n\log n$ for $n>2$ gives the estimate for ϕ.

6 Average orders

It is often of interest to determine the magnitude 'on average' of arithmetical functions f, that is, to find estimates for sums of the form $\sum f(n)$ with $n\le x$, where x is a large real number. We shall obtain such estimates when f is τ, σ and ϕ.

First we observe that

$$\sum_{n\le x}\tau(n)=\sum_{n\le x}\sum_{d|n}1=\sum_{d\le x}\sum_{m\le x/d}1=\sum_{d\le x}[x/d].$$

Now we have

$$\sum_{d\le x}1/d=\log x+O(1),$$

and hence

$$\sum_{n\le x} \tau(n) = x \log x + O(x).$$

This implies that $(1/x)\sum \tau(n) \sim \log x$ as $x \to \infty$. The argument can be refined to give

$$\sum_{n\le x} \tau(n) = x \log x + (2\gamma - 1)x + O(\sqrt{x}),$$

where γ is Euler's constant. Note that although one can say that the 'average order' of $\tau(n)$ is $\log n$ (since $\sum \log n \sim x \log x$), it is not true that 'almost all' numbers have about $\log n$ divisors; here almost all numbers are said to have a certain property if the proportion $\le x$ not possessing the property is $o(x)$. In fact 'almost all' numbers have about $(\log n)^{\log 2}$ divisors, that is, for any $\varepsilon > 0$ and for almost all n, the function $\tau(n)/(\log n)^{\log 2}$ lies between $(\log n)^{\varepsilon}$ and $(\log n)^{-\varepsilon}$.

To determine the average order of $\sigma(n)$ we observe that

$$\sum_{n\le x} \sigma(n) = \sum_{n\le x}\sum_{d|n} (n/d) = \sum_{d\le x}\sum_{m\le x/d} m.$$

The last sum is

$$\tfrac{1}{2}[x/d]([x/d]+1) = \tfrac{1}{2}(x/d)^2 + O(x/d).$$

Now

$$\sum_{d\le x} 1/d^2 = \sum_{d=1}^{\infty} 1/d^2 + O(1/x),$$

and thus we obtain

$$\sum_{n\le x} \sigma(n) = \frac{1}{12}\pi^2 x^2 + O(x \log x).$$

This implies that the 'average order' of $\sigma(n)$ is $\frac{1}{6}\pi^2 n$ (since $\sum n \sim \frac{1}{2}x^2$).

Finally we derive an average estimate for ϕ. We have

$$\sum_{n\le x} \phi(n) = \sum_{n\le x}\sum_{d|n} \mu(d)(n/d) = \sum_{d\le x} \mu(d) \sum_{m\le x/d} m.$$

The last sum is

$$\tfrac{1}{2}(x/d)^2 + O(x/d).$$

Now

$$\sum_{d\le x} \mu(d)/d^2 = \sum_{d=1}^{\infty} \mu(d)/d^2 + O(1/x),$$

and the infinite series here has sum $6/\pi^2$, as will be clear from

§ 8. Hence we obtain

$$\sum_{n\le x}\phi(n)=(3/\pi^2)x^2+O(x\log x).$$

This implies that the 'average order' of $\phi(n)$ is $6n/\pi^2$. Moreover the result shows that the probability that two integers be relatively prime is $6/\pi^2$. For there are $\frac{1}{2}n(n+1)$ pairs of integers p, q with $1\le p\le q\le n$, and precisely $\phi(1)+\cdots+\phi(n)$ of the corresponding fractions p/q are in their lowest terms.

7 Perfect numbers

A natural number n is said to be perfect if $\sigma(n)=2n$, that is if n is equal to the sum of its divisors other than itself. Thus, for instance, 6 and 28 are perfect numbers.

Whether there exist any odd perfect numbers is a notorious unresolved problem. By contrast, however, the even perfect numbers can be specified precisely. Indeed an even number is perfect if and only if it has the form $2^{p-1}(2^p-1)$, where both p and 2^p-1 are primes. It suffices to prove the necessity, for it is readily verified that numbers of this form are certainly perfect. Suppose therefore that $\sigma(n)=2n$ and that $n=2^km$, where k and m are positive integers with m odd. We have $(2^{k+1}-1)\sigma(m)=2^{k+1}m$ and hence $\sigma(m)=2^{k+1}l$ and $m=(2^{k+1}-1)l$ for some positive integer l. If now l were greater than 1 then m would have distinct divisors l, m and 1, whence we would have $\sigma(m)\ge l+m+1$. But $l+m=2^{k+1}l=\sigma(m)$, and this gives a contradiction. Thus $l=1$ and $\sigma(m)=m+1$, which implies that m is a prime. In fact m is a Mersenne prime and hence $k+1$ is a prime p, say (cf. § 6 of Chapter 1). This shows that n has the required form.

8 The Riemann zeta-function

In a classic memoir of 1860 Riemann showed that questions concerning the distribution of the primes are intimately related to properties of the zeta-function

$$\zeta(s)=\sum_{n=1}^{\infty}1/n^s,$$

where s denotes a complex variable. It is clear that the series converges absolutely for $\sigma>1$, where $s=\sigma+it$ with σ, t real, and indeed that it converges uniformly for $\sigma>1+\delta$ for any

$\delta > 0$. Riemann showed that $\zeta(s)$ can be continued analytically throughout the complex plane and that it is regular there except for a simple pole at $s = 1$ with residue 1. He showed moreover that it satisfies the functional equation $\Xi(s) = \Xi(1 - s)$, where

$$\Xi(s) = \pi^{-\frac{1}{2}s}\Gamma(\tfrac{1}{2}s)\zeta(s).$$

The fundamental connection between the zeta-function and the primes is given by the Euler product

$$\zeta(s) = \prod_p (1 - 1/p^s)^{-1},$$

valid for $\sigma > 1$. The relation is readily verified; in fact it is clear that, for any positive integer N,

$$\prod_{p \le N} (1 - 1/p^s)^{-1} = \prod_{p \le N} (1 + p^{-s} + p^{-2s} + \cdots) = \sum_m m^{-s},$$

where m runs through all the positive integers that are divisible only by primes $\le N$, and

$$\left| \sum_m m^{-s} - \sum_{n \le N} n^{-s} \right| \le \sum_{n > N} n^{-\sigma} \to 0 \quad \text{as } N \to \infty.$$

The Euler product shows that $\zeta(s)$ has no zeros for $\sigma > 1$. In view of the functional equation it follows that $\zeta(s)$ has no zeros for $\sigma < 0$ except at the points $s = -2, -4, -6, \ldots$; these are termed the 'trivial zeros'. All other zeros of $\zeta(s)$ must lie in the 'critical strip' given by $0 \le \sigma \le 1$, and Riemann conjectured that they in fact lie on the line $\sigma = \frac{1}{2}$. This is the famous Riemann hypothesis and it remains unproved to this day. There is much evidence in favour of the hypothesis; in particular Hardy proved in 1915 that infinitely many zeros of $\zeta(s)$ lie on the critical line, and extensive computations have verified that at least the first three million zeros above the real axis do so. It has been shown that, if the hypothesis is true, then, for instance, there is a refinement of the prime number theorem to the effect that

$$\pi(x) = \int_2^x \frac{dt}{\log t} + O(\sqrt{x} \log x),$$

and that the difference between consecutive primes satisfies $p_{n+1} - p_n = O(p_n^{\frac{1}{2}+\varepsilon})$. In fact it has been shown that there is a narrow zero-free region for $\zeta(s)$ to the left of the line $\sigma = 1$, and this implies that results as above are indeed valid but with weaker error terms. It is also known that the Riemann hypothesis is equivalent to the assertion that, for any $\varepsilon > 0$,

$$\sum_{n \leq x} \mu(n) = O(x^{\frac{1}{2}+\varepsilon}).$$

The basic relation between the Möbius function and the Riemann zeta-function is given by

$$1/\zeta(s) = \sum_{n=1}^{\infty} \mu(n)/n^s.$$

This is clearly valid for $\sigma > 1$ since the product of the series on the right with $\sum 1/n^s$ is $\sum \nu(n)/n^s$. In fact if the Riemann hypothesis holds then the equation remains true for $\sigma > \frac{1}{2}$. There is a similar equation for the Euler function, valid for $\sigma > 2$, namely

$$\zeta(s-1)/\zeta(s) = \sum_{n=1}^{\infty} \phi(n)/n^s.$$

This is readily verified from the result at the end of § 3. Likewise there are equations for $\tau(n)$ and $\sigma(n)$, valid respectively for $\sigma > 1$ and $\sigma > 2$, namely

$$(\zeta(s))^2 = \sum_{n=1}^{\infty} \tau(n)/n^s, \qquad \zeta(s)\zeta(s-1) = \sum_{n=1}^{\infty} \sigma(n)/n^s.$$

9 Further reading

The elementary arithmetical functions are discussed in every introductory text on number theory; again Hardy and Wright is a good reference. As regards the last section, the most comprehensive work on the subject is that of E. C. Titchmarsh *The theory of the Riemann zeta-function* (Oxford U.P., 1951). Other books to be recommended are those of T. M. Apostol (Springer-Verlag, Berlin, 1976) and K. Chandrasekharan (Springer-Verlag, Berlin, 1968), both with the title *Introduction to analytic number theory*; see also Chandrasekharan's *Arithmetical functions* (Springer-Verlag, Berlin, 1970).

10 Exercises

(i) Evaluate $\sum_{d|n} \mu(d)\sigma(d)$ in terms of the distinct prime factors of n.

(ii) Let $\Lambda(n) = \log p$ if n is a power of a prime p and let $\Lambda(n) = 0$ otherwise (Λ is called von Mangoldt's function). Evaluate $\sum_{d|n} \Lambda(d)$. Express $\sum \Lambda(n)/n^s$ in terms of $\zeta(s)$.

(iii) Let a run through all the integers with $1 \leq a \leq n$ and $(a, n) = 1$. Show that $f(n) = (1/n) \sum a$ satisfies $\sum_{d|n} f(d) = \frac{1}{2}(n+1)$. Hence prove that $f(n) = \frac{1}{2}\phi(n)$ for $n > 1$.

(iv) Let a run through the integers as in (iii). Prove that $(1/n^3) \sum a^3 = \frac{1}{4}\phi(n)(1 + (-1)^k p_1 \cdots p_k/n^2)$, where $p_1, \ldots, p_k$ are the distinct prime factors of n (>1).

(v) Show that the product of all the integers a in (iii) is given by $n^{\phi(n)} \prod_{d|n} (d!/d^d)^{\mu(n/d)}$.

(vi) Show that $\sum_{n \leq x} \mu(n)[x/n] = 1$. Hence prove that $|\sum_{n \leq x} \mu(n)/n| \leq 1$.

(vii) Let m, n be positive integers and let d run through all divisors of (m, n). Prove that $\sum d\mu(n/d) = \mu(n/(m, n))\phi(n)/\phi(n/(m, n))$. (The sum here is called Ramanujan's sum.)

(viii) Prove that $\sum_{n=1}^{\infty} \phi(n)x^n/(1 - x^n) = x/(1 - x)^2$. (Series of this kind are called Lambert series.)

(ix) Prove that $\sum_{n \leq x} \phi(n)/n = (6/\pi^2)x + O(\log x)$.

3

Congruences

1 Definitions

Suppose that a, b are integers and that n is a natural number. By $a \equiv b \pmod{n}$ one means n divides $b-a$; and one says that a is congruent to b modulo n. If $0 \leq b < n$ then one refers to b as the residue of $a \pmod{n}$. It is readily verified that the congruence relation is an equivalence relation; the equivalence classes are called residue classes or congruence classes. By a complete set of residues $(\text{mod } n)$ one means a set of n integers one from each residue class $(\text{mod } n)$.

It is clear that if $a \equiv a' \pmod{n}$ and $b \equiv b' \pmod{n}$ then $a+b \equiv a'+b'$ and $a-b \equiv a'-b' \pmod{n}$. Further we have $ab \equiv a'b' \pmod{n}$, since n divides $(a-a')b + a'(b-b')$. Furthermore, if $f(x)$ is any polynomial with integer coefficients, then $f(a) \equiv f(a') \pmod{n}$.

Note also that if $ka \equiv ka' \pmod{n}$ for some natural number k with $(k, n) = 1$ then $a \equiv a' \pmod{n}$: thus if $a_1, \ldots, a_n$ is a complete set of residues $(\text{mod } n)$ then so is $ka_1, \ldots, ka_n$. More generally, if k is any natural number such that $ka \equiv ka' \pmod{n}$ then $a \equiv a' \pmod{n/(k, n)}$, since obviously $k/(k, n)$ and $n/(k, n)$ are relatively prime.

2 Chinese remainder theorem

Let a, n be natural numbers and let b be any integer. We prove first that the linear congruence $ax \equiv b \pmod{n}$ is soluble for some integer x if and only if (a, n) divides b. The condition is certainly necessary, for (a, n) divides both a and n. To prove the sufficiency, suppose that $d = (a, n)$ divides b. Put $a' = a/d$, $b' = b/d$ and $n' = n/d$. Then it suffices to solve $a'x \equiv b' \pmod{n'}$. But this has precisely one solution $(\text{mod } n')$, since $(a', n') = 1$ and so $a'x$ runs through a complete set of residues (mod n') as x runs through such a set. It is clear that if x' is any solution of $a'x' \equiv b' \pmod{n'}$ then the complete set of solutions (mod n) of $ax \equiv b \pmod{n}$ is given by $x = x' + mn'$, where $m = 1, 2, \ldots, d$. Hence, when d divides b, the congruence $ax \equiv b \pmod{n}$ has precisely d solutions (mod n).

It follows from the last result that if p is a prime and if a is not divisible by p then the congruence $ax \equiv b \pmod{p}$ is always soluble; in fact there is a unique solution (mod p). This implies that the residues $0, 1, \ldots, p-1$ form a field under addition and multiplication (mod p). It is usual to denote the field by $\mathbb{Z}_p$.

We turn now to simultaneous linear congruences and prove the Chinese remainder theorem; the result was apparently known to the Chinese at least 1500 years ago. Let $n_1, \ldots, n_k$ be natural numbers and suppose that they are coprime in pairs, that is $(n_i, n_j) = 1$ for $i \neq j$. The theorem asserts that, for any integers $c_1, \ldots, c_k$, the congruences $x \equiv c_j \pmod{n_j}$, with $1 \leq j \leq k$, are soluble simultaneously for some integer x; in fact there is a unique solution modulo $n = n_1 \cdots n_k$. For the proof, let $m_j = n/n_j$ $(1 \leq j \leq k)$. Then $(m_j, n_j) = 1$ and thus there is an integer x_j such that $m_j x_j \equiv c_j \pmod{n_j}$. Now it is readily seen that $x = m_1 x_1 + \cdots + m_k x_k$ satisfies $x \equiv c_j \pmod{n_j}$, as required. The uniqueness is clear, for if x, y are two solutions then $x \equiv y \pmod{n_j}$ for $1 \leq j \leq k$, whence, since the n_j are coprime in pairs, we have $x \equiv y \pmod{n}$. Plainly the Chinese remainder theorem together with the first result of this section implies that if $n_1, \ldots, n_k$ are coprime in pairs then the congruences $a_j x \equiv b_j \pmod{n_j}$, with $1 \leq j \leq k$, are soluble simultaneously if and only if (a_j, n_j) divides b_j for all j.

As an example, consider the congruences $x \equiv 2 \pmod{5}$, $x \equiv 3 \pmod{7}$, $x \equiv 4 \pmod{11}$. In this case a solution is given by $x = 77x_1 + 55x_2 + 35x_3$, where x_1, x_2, x_3 satisfy $2x_1 \equiv 2 \pmod{5}$, $6x_2 \equiv 3 \pmod{7}$, $2x_3 \equiv 4 \pmod{11}$. Thus we can take $x_1 = 1$, $x_2 = 4$, $x_3 = 2$, and these give $x = 367$. The complete solution is $x \equiv -18 \pmod{385}$.

3 The theorems of Fermat and Euler

First we introduce the concept of a reduced set of residues (mod n). By this we mean a set of $\phi(n)$ numbers one from each of the $\phi(n)$ residue classes that consist of numbers relatively prime to n. In particular, the numbers a with $1 \le a \le n$ and $(a, n) = 1$ form a reduced set of residues (mod n).

We proceed now to establish the multiplicative property of ϕ, referred to in § 3 of Chapter 2, using the above concept. Accordingly let n, n' be natural numbers with $(n, n') = 1$. Further let a and a' run through reduced sets of residues (mod n) and (mod n') respectively. Then it suffices to prove that $an' + a'n$ runs through a reduced set of residues (mod nn'); for this implies that $\phi(n)\phi(n') = \phi(nn')$, as required. Now clearly, since $(a, n) = 1$ and $(a', n') = 1$, the number $an' + a'n$ is relatively prime to n and to n' and so to nn'. Furthermore any two distinct numbers of the form are incongruent (mod nn'). Thus we have only to prove that if $(b, nn') = 1$ then $b \equiv an' + a'n$ (mod nn') for some a, a' as above. But since $(n, n') = 1$ there exist integers m, m' satisfying $mn' + m'n = 1$. Plainly $(bm, n) = 1$ and so $a \equiv bm$ (mod n) for some a; similary $a' \equiv bm'$ (mod n') for some a', and now it is easily seen that a, a' have the required property.

Fermat's theorem states that if a is any natural number and if p is any prime then $a^p \equiv a$ (mod p). In particular, if $(a, p) = 1$, then $a^{p-1} \equiv 1$ (mod p). The theorem was announced by Fermat in 1640 but without proof. Euler gave the first demonstration about a century later and, in 1760, he established a more general result to the effect that, if a, n are natural numbers with $(a, n) = 1$, then $a^{\phi(n)} \equiv 1$ (mod n). For the proof of Euler's theorem, we observe simply that as x runs through a reduced set of residues (mod n) so also ax runs through such a set. Hence $\prod (ax) \equiv \prod (x)$ (mod n), where the products are taken over all x in the reduced set, and the theorem follows on cancelling $\prod (x)$ from both sides.

4 Wilson's theorem

This asserts that $(p-1)! \equiv -1$ (mod p) for any prime p. Though the result is attributed to Wilson, the statement was apparently first published by Waring in his *Meditationes algebraicae* of 1770 and a proof was furnished a little later by Lagrange.

For the demonstration, it suffices to assume that p is odd. Now to every integer a with $0 < a < p$ there is a unique integer a'

with $0<a'<p$ such that $aa'\equiv 1 \pmod p$. Further, if $a=a'$ then $a^2\equiv 1 \pmod p$ whence $a=1$ or $a=p-1$. Thus the set $2, 3, \ldots, p-2$ can be divided into $\frac{1}{2}(p-3)$ pairs a, a' with $aa'\equiv 1 \pmod p$. Hence we have $2\cdot 3\cdots(p-2)\equiv 1 \pmod p$, and so $(p-1)!\equiv p-1\equiv -1 \pmod p$, as required.

Wilson's theorem admits a converse and so yields a criterion for primes. Indeed an integer $n>1$ is a prime if and only if $(n-1)!\equiv -1 \pmod n$. To verify the sufficiency note that any divisor of n, other than itself, must divide $(n-1)!$.

As an immediate deduction from Wilson's theorem we see that if p is a prime with $p\equiv 1 \pmod 4$ then the congruence $x^2\equiv -1 \pmod p$ has solutions $x=\pm(r!)$, where $r=\frac{1}{2}(p-1)$. This follows on replacing $a+r$ in $(p-1)!$ by the congruent integer $a-r-1$ for each a with $1\le a\le r$. Note that the congruence has no solutions when $p\equiv 3 \pmod 4$, for otherwise we would have $x^{p-1}=x^{2r}\equiv(-1)^r=-1 \pmod p$ contrary to Fermat's theorem.

5 Lagrange's theorem

Let $f(x)$ be a polynomial with integer coefficients and with degree n. Suppose that p is a prime and that the leading coefficient of f, that is the coefficient of x^n, is not divisible by p. Lagrange's theorem states that the congruence $f(x)\equiv 0 \pmod p$ has at most n solutions $\pmod p$.

The theorem certainly holds for $n=1$ by the first result in § 2. We assume that it is valid for polynomials with degree $n-1$ and proceed inductively to prove the theorem for polynomials with degree n. Now, for any integer a we have $f(x)-f(a)=(x-a)g(x)$, where g is a polynomial with degree $n-1$, with integer coefficients and with the same leading coefficient as f. Thus if $f(x)\equiv 0 \pmod p$ has a solution $x=a$ then all solutions of the congruence satisfy $(x-a)g(x)\equiv 0 \pmod p$. But, by the inductive hypothesis, the congruence $g(x)\equiv 0 \pmod p$ has at most $n-1$ solutions $\pmod p$. The theorem follows. It is customary to write $f(x)\equiv g(x) \pmod p$ to signify that the coefficients of like powers of x in the polynomials f, g are congruent $\pmod p$; and it is clear that if the congruence $f(x)\equiv 0 \pmod p$ has its full complement $a_1, \ldots, a_n$ of solutions $\pmod p$ then

$$f(x)\equiv c(x-a_1)\cdots(x-a_n) \pmod p,$$

where c is the leading coefficient of f. In particular, by Fermat's theorem, we have

$$x^{p-1}-1\equiv(x-1)\cdots(x-p+1)(\bmod p),$$

and, on comparing constant coefficients, we obtain another proof of Wilson's theorem.

Plainly, instead of speaking of congruences, we can express the above succinctly in terms of polynomials defined over $\mathbb{Z}_p$. Thus Lagrange's theorem asserts that the number of zeros in $\mathbb{Z}_p$ of a polynomial defined over this field cannot exceed its degree. As a corollary we deduce that, if d divides $p-1$ then the polynomial x^d-1 has precisely d zeros in $\mathbb{Z}_p$. For we have $x^{p-1}-1=(x^d-1)g(x)$, where g has degree $p-1-d$. But, by Fermat's theorem, $x^{p-1}-1$ has $p-1$ zeros in $\mathbb{Z}_p$ and so x^d-1 has at least $(p-1)-(p-1-d)=d$ zeros in $\mathbb{Z}_p$, whence the assertion.

Lagrange's theorem does not remain true for composite moduli. In fact it is readily verified from the Chinese remainder theorem that if $m_1,\ldots,m_k$ are natural numbers coprime in pairs, if $f(x)$ is a polynomial with integer coefficients, and if the congruence $f(x)\equiv 0\ (\bmod m_j)$ has s_j solutions $(\bmod m_j)$, then the congruence $f(x)\equiv 0\ (\bmod m)$, where $m=m_1\cdots m_k$, has $s=s_1\cdots s_k$ solutions $(\bmod m)$. Lagrange's theorem is still false for prime power moduli; for example $x^2\equiv 1\ (\bmod 8)$ has four solutions. But if the prime p does not divide the discriminant of f then the theorem holds for all powers p^j; indeed the number of solutions of $f(x)\equiv 0\ (\bmod p^j)$ is, in this case, the same as the number of solutions of $f(x)\equiv 0\ (\bmod p)$. This can be seen at once when, for instance, $f(x)=x^2-a$; for if p is any odd prime that does not divide a, then from a solution y of $f(y)\equiv 0\ (\bmod p^j)$ we obtain a solution $x=y+p^jz$ of $f(x)\equiv 0\ (\bmod p^{j+1})$ by solving the congruence $2yz+f(y)/p^j\equiv 0\ (\bmod p)$ for z, as is possible since $(2y,p)=1$.

6 Primitive roots

Let a, n be natural numbers with $(a,n)=1$. The least natural number d such that $a^d\equiv 1\ (\bmod n)$ is called the order of $a\ (\bmod n)$, and a is said to belong to $d\ (\bmod n)$. By Euler's theorem, the order d exists and it divides $\phi(n)$. In fact d divides every integer k such that $a^k \equiv 1 \pmod{n}$, for, by the division algorithm, $k = dq + r$ with $0 \le r < d$, whence $a^r \equiv 1 \pmod{n}$ and so $r = 0$.

By a primitive root $(\bmod n)$ we mean a number that belongs to $\phi(n) \pmod{n}$. We proceed to prove that for every odd prime p there exist $\phi(p-1)$ primitive roots $(\bmod p)$. Now each of the numbers $1, 2, \ldots, p-1$ belongs $(\bmod p)$ to some divisor d of $p-1$; let $\psi(d)$ be the number that belongs to $d \pmod{p}$ so that

$$\sum_{d|(p-1)} \psi(d) = p - 1.$$

It will suffice to prove that if $\psi(d) \neq 0$ then $\psi(d) = \phi(d)$. For, by § 3 of Chapter 2, we have

$$\sum_{d|(p-1)} \phi(d) = p - 1,$$

whence $\psi(d) \neq 0$ for all d and so $\psi(p-1) = \phi(p-1)$ as required.

To verify the assertion concerning ψ, suppose that $\psi(d) \neq 0$ and let a be a number that belongs to $d \pmod{p}$. Then $a, a^2, \ldots, a^d$ are mutually incongruent solutions of $x^d \equiv 1 \pmod{p}$ and thus, by Lagrange's theorem, they represent all the solutions (in fact we showed in § 5 that the congruence has precisely d solutions $(\bmod p)$). It is now easily seen that the numbers a^m with $1 \le m \le d$ and $(m, d) = 1$ represent all the numbers that belong to $d \pmod{p}$; indeed each has order d, for if $a^{md'} \equiv 1$ then $d|d'$, and if b is any number that belongs to $d \pmod{p}$ then $b \equiv a^m$ for some m with $1 \le m \le d$, and we have $(m, d) = 1$ since $b^{d/(m,d)} \equiv (a^d)^{m/(m,d)} \equiv 1 \pmod{p}$. This gives $\psi(d) = \phi(d)$, as asserted.

Let g be a primitive root $(\bmod p)$. We prove now that there exists an integer x such that $g' = g + px$ is a primitive root $(\bmod p^j)$ for all prime powers p^j. We have $g^{p-1} = 1 + py$ for some integer y and so, by the binomial theorem, $g'^{p-1} = 1 + pz$, where

$$z \equiv y + (p-1)g^{p-2}x \pmod{p}.$$

The coefficient of x is not divisible by p and so we can choose x such that $(z, p) = 1$. Then g' has the required property. For suppose that g' belongs to $d \pmod{p^j}$. Then d divides $\phi(p^j) = p^{j-1}(p-1)$. But g' is a primitive root $(\bmod p)$ and thus $p-1$ divides d. Hence $d = p^k(p-1)$ for some $k < j$. Further, since p

is odd, we have

$$(1+pz)^{p^k}=1+p^{k+1}z_k,$$

where $(z_k, p)=1$. Now since $g'^d \equiv 1 \pmod{p^j}$ it follows that $j=k+1$ and this gives $d=\phi(p^j)$, as required.

Finally we deduce that, for any natural number n, there exists a primitive root (mod n) if and only if n has the form 2, 4, p^j or $2p^j$, where p is an odd prime. Clearly 1 and 3 are primitive roots (mod 2) and (mod 4). Further, if g is a primitive root (mod p^j) then the odd element of the pair g, $g+p^j$ is a primitive root (mod $2p^j$), since $\phi(2p^j)=\phi(p^j)$. Hence it remains only to prove the necessity of the assertion. Now if $n=n_1n_2$, where $(n_1, n_2)=1$ and $n_1>2$, $n_2>2$, then there is no primitive root (mod n). For $\phi(n_1)$ and $\phi(n_2)$ are even and thus for any natural number a we have

$$a^{\frac{1}{2}\phi(n)}=(a^{\phi(n_1)})^{\frac{1}{2}\phi(n_2)}\equiv 1 \pmod{n_1};$$

similarly $a^{\frac{1}{2}\phi(n)}\equiv 1 \pmod{n_2}$, whence $a^{\frac{1}{2}\phi(n)}\equiv 1 \pmod{n}$. Further, there are no primitive roots (mod 2^j) for $j>2$, since, by induction, we have $a^{2^{j-2}}\equiv 1 \pmod{2^j}$ for all odd numbers a. This proves the theorem.

7 Indices

Let g be a primitive root (mod n). The numbers g^l with $l=0, 1, \ldots, \phi(n)-1$ form a reduced set of residues (mod n). Hence, for every integer a with $(a, n)=1$ there is a unique l such that $g^l \equiv a \pmod{n}$. The exponent l is called the index of a with respect to g and it is denoted by ind a. Plainly we have

$$\text{ind } a+\text{ind } b \equiv \text{ind }(ab) \pmod{\phi(n)},$$

and ind $1=0$, ind $g=1$. Further, for every natural number m, we have $\text{ind }(a^m)\equiv m \text{ ind } a \pmod{\phi(n)}$. These properties of the index are clearly analogous to the properties of logarithms. We also have $\text{ind }(-1)=\frac{1}{2}\phi(n)$ for $n>2$ since $g^{2 \text{ ind}(-1)}\equiv 1 \pmod{n}$ and $2 \text{ ind }(-1)<2\phi(n)$.

As an example of the use of indices, consider the congruence $x^n\equiv a \pmod{p}$, where p is a prime. We have $n \text{ ind } x \equiv \text{ind } a \pmod{(p-1)}$ and thus if $(n, p-1)=1$ then there is just one solution. Consider, in particular, $x^5\equiv 2 \pmod{7}$. It is readily verified that 3 is a primitive root (mod 7) and we have $3^2 \equiv 2 \pmod{7}$. Thus $5 \operatorname{ind} x \equiv 2 \pmod{6}$, which gives $\operatorname{ind} x = 4$ and $x \equiv 3^4 \equiv 4 \pmod{7}$.

Note that although there is no primitive root (mod 2^j) for $j > 2$, the number 5 belongs to 2^{j-2} (mod 2^j) and every odd integer a is congruent (mod 2^j) to just one integer of the form $(-1)^l 5^m$, where $l = 0, 1$ and $m = 0, 1, \ldots, 2^{j-2} - 1$. The pair l, m has similar properties to the index defined above.

8 Further reading

A good account of the elementary theory of congruences is given by T. Nagell, *Introduction to number theory* (Wiley, New York, 1951); this contains, in particular, a table of primitive roots. There is another, and in fact more extensive table in I. M. Vinogradov's *An introduction to the theory of numbers* (Pergamon Press, Oxford, London, New York, Paris, 1961). Again Hardy and Wright cover the subject well.

9 Exercises

(i) Find an integer x such that $2x \equiv 1 \pmod{3}$, $3x \equiv 1 \pmod{5}$, $5x \equiv 1 \pmod{7}$.

(ii) Prove that for any positive integers a, n with $(a, n) = 1$, $\sum \{ax/n\} = \frac{1}{2}\phi(n)$, where the summation is over all x in a reduced set of residues (mod n).

(iii) The integers a and $n > 1$ satisfy $a^{n-1} \equiv 1 \pmod{n}$ but $a^m \not\equiv 1 \pmod{n}$ for each divisor m of $n-1$, other than itself. Prove that n is a prime.

(iv) Show that the congruence $x^{p-1} - 1 \equiv 0 \pmod{p^j}$ has just $p-1$ solutions (mod p^j) for every prime power p^j.

(v) Prove that, for every natural number n, either there is no primitive root (mod n) or there are $\phi(\phi(n))$ primitive roots (mod n).

(vi) Prove that, for any prime p, the sum of all the distinct primitive roots (mod p) is congruent to $\mu(p-1)$ (mod p).

(vii) Determine all the solutions of the congruence $y^2 \equiv 5x^3 \pmod 7$ in integers x, y.

(viii) Prove that if p is a prime >3 then the numerator of $1+\frac{1}{2}+\cdots+1/(p-1)$ is divisible by p^2 (Wolstenholme's theorem).

4

Quadratic residues

1 Legendre's symbol

In the last chapter we discussed the linear congruence $ax \equiv b \pmod{n}$. Here we shall study the quadratic congruence $x^2 \equiv a \pmod{n}$; in fact this amounts to the study of the general quadratic congruence $ax^2 + bx + c \equiv 0 \pmod{n}$, since on writing $d = b^2 - 4ac$ and $y = 2ax + b$, the latter gives $y^2 \equiv d \pmod{4an}$.

Let a be any integer, let n be a natural number and suppose that $(a, n) = 1$. Then a is called a quadratic residue (mod n) if the congruence $x^2 \equiv a \pmod{n}$ is soluble; otherwise it is called a quadratic non-residue (mod n). The Legendre symbol $\left(\frac{a}{p}\right)$, where p is a prime and $(a, p) = 1$, is defined as 1 if a is a quadratic residue (mod p) and as -1 if a is a quadratic non-residue (mod p). Clearly, if $a \equiv a' \pmod{p}$, we have

$$\left(\frac{a}{p}\right) = \left(\frac{a'}{p}\right).$$

2 Euler's criterion

This states that if p is an odd prime then

$$\left(\frac{a}{p}\right) \equiv a^{\frac{1}{2}(p-1)} \pmod{p}.$$

For the proof we write, for brevity, $r = \frac{1}{2}(p-1)$ and we note first that if a is a quadratic residue (mod p) then for some x in $\mathbb{N}$ we have $x^2 \equiv a \pmod{p}$, whence, by Fermat's theorem, $a^r \equiv x^{p-1} \equiv 1 \pmod{p}$. Thus it suffices to show that if a is a quadratic non-residue (mod p) then $a^r \equiv -1 \pmod{p}$. Now in any reduced set of residues (mod p) there are r quadratic residues (mod p)

and r quadratic non-residues (mod p); for the numbers $1^2, 2^2, \ldots, r^2$ are mutually incongruent (mod p) and since, for any integer k, $(p-k)^2 \equiv k^2 \pmod{p}$, the numbers represent all the quadratic residues (mod p). Each of the numbers satisfies $x^r \equiv 1 \pmod{p}$, and, by Lagrange's theorem, the congruence has at most r solutions (mod p). Hence if a is a quadratic non-residue (mod p) then a is not a solution of the congruence. But, by Fermat's theorem, $a^{p-1} \equiv 1 \pmod{p}$, whence $a^r \equiv \pm 1 \pmod{p}$. The required result follows. Note that one can argue alternatively in terms of a primitive root (mod p), say g; indeed it is clear that the quadratic residues (mod p) are given by $1, g^2, \ldots, g^{2r}$.

As an immediate corollary to Euler's criterion we have the multiplicative property of the Legendre symbol, namely

$$\left(\frac{a}{p}\right)\left(\frac{b}{p}\right) = \left(\frac{ab}{p}\right)$$

for all integers a, b not divisible by p; here equality holds since both sides are ± 1. Similarly we have

$$\left(\frac{-1}{p}\right) = (-1)^{\frac{1}{2}(p-1)};$$

in other words, -1 is a quadratic residue of all primes $\equiv 1 \pmod{4}$ and a quadratic non-residue of all primes $\equiv 3 \pmod{4}$. It will be recalled from §4 of Chapter 3 that when $p \equiv 1 \pmod{4}$ the solutions of $x^2 \equiv -1 \pmod{p}$ are given by $x = \pm(r!)$.

3 Gauss' lemma

For any integer a and any natural number n we define the numerically least residue of a (mod n) as that integer a' for which $a \equiv a' \pmod{n}$ and $-\frac{1}{2}n < a' \leq \frac{1}{2}n$.

Let now p be an odd prime and suppose that $(a, p) = 1$. Further let a_j be the numerically least residue of aj (mod p) for $j = 1, 2, \ldots$. Then Gauss' lemma states that

$$\left(\frac{a}{p}\right) = (-1)^l,$$

where l is the number of $j \leq \frac{1}{2}(p-1)$ for which $a_j < 0$.

For the proof we observe that the numbers $|a_j|$ with $1 \leq j \leq r$, where $r = \frac{1}{2}(p-1)$, are simply the numbers $1, 2, \ldots, r$ in some order. For certainly we have $1 \le |a_j| \le r$, and the $|a_j|$ are distinct since $a_j = -a_k$, with $k \le r$, would give $a(j+k) \equiv 0 \pmod p$ with $0 < j+k < p$, which is impossible, and $a_j = a_k$ gives $aj \equiv ak \pmod p$, whence $j = k$. Hence we have $a_1 \cdots a_r = (-1)^l r!$. But $a_j \equiv aj \pmod p$ and so $a_1 \cdots a_r \equiv a^r r! \pmod p$. Thus $a^r \equiv (-1)^l \pmod p$, and the result now follows from Euler's criterion.

As a corollary we obtain

$$\left(\frac{2}{p}\right) = (-1)^{\frac{1}{8}(p^2-1)},$$

that is, 2 is a quadratic residue of all primes $\equiv \pm 1 \pmod 8$ and a quadratic non-residue of all primes $\equiv \pm 3 \pmod 8$. To verify this result, note that, when $a = 2$, we have $a_j = 2j$ for $1 \le j \le [\frac{1}{4}p]$ and $a_j = 2j - p$ for $[\frac{1}{4}p] < j \le \frac{1}{2}(p-1)$. Hence in this case $l = \frac{1}{2}(p-1) - [\frac{1}{4}p]$, and it is readily checked that $l \equiv \frac{1}{8}(p^2-1) \pmod 2$.

4 Law of quadratic reciprocity

We come now to the famous theorem stated by Euler in 1783 and first proved by Gauss in 1796. Apparently Euler, Legendre and Gauss each discovered the theorem independently and Gauss worked on it intensively for a year before establishing the result; he subsequently gave no fewer than eight demonstrations.

The law of quadratic reciprocity asserts that if p, q are distinct odd primes then

$$\left(\frac{p}{q}\right)\left(\frac{q}{p}\right) = (-1)^{\frac{1}{4}(p-1)(q-1)}.$$

Thus if p, q are not both congruent to $3 \pmod 4$ then

$$\left(\frac{p}{q}\right) = \left(\frac{q}{p}\right),$$

and in the exceptional case

$$\left(\frac{p}{q}\right) = -\left(\frac{q}{p}\right).$$

For the proof we observe that, by Gauss' lemma, $\left(\frac{p}{q}\right) = (-1)^l$, where l is the number of lattice points (x, y) (that is, pairs of integers) satisfying $0 < x < \frac{1}{2}q$ and $-\frac{1}{2}q < px - qy < 0$. Now these inequalities give $y<(px/q)+\frac{1}{2}<\frac{1}{2}(p+1)$. Hence, since y is an integer, we see that l is the number of lattice points in the rectangle R defined by $0<x<\frac{1}{2}q$, $0<y<\frac{1}{2}p$, satisfying $-\frac{1}{2}q<px-qy<0$ (see Fig. 4.1). Similarly

$$\left(\frac{q}{p}\right)=(-1)^m,$$

where m is the number of lattice points in R satisfying $-\frac{1}{2}p<qy-px<0$. Now it suffices to prove that $\frac{1}{4}(p-1)(q-1)-(l+m)$ is even. But $\frac{1}{4}(p-1)(q-1)$ is just the number of lattice points in R, and thus the latter expression is the number of lattice points in R satisfying either $px-qy\leq-\frac{1}{2}q$ or $qy-px\leq-\frac{1}{2}p$. The regions in R defined by these inequalities are disjoint and they contain the same number of lattice points since, as is readily verified, the substitution

$$x=\tfrac{1}{2}(q+1)-x', \; y=\tfrac{1}{2}(p+1)-y'$$

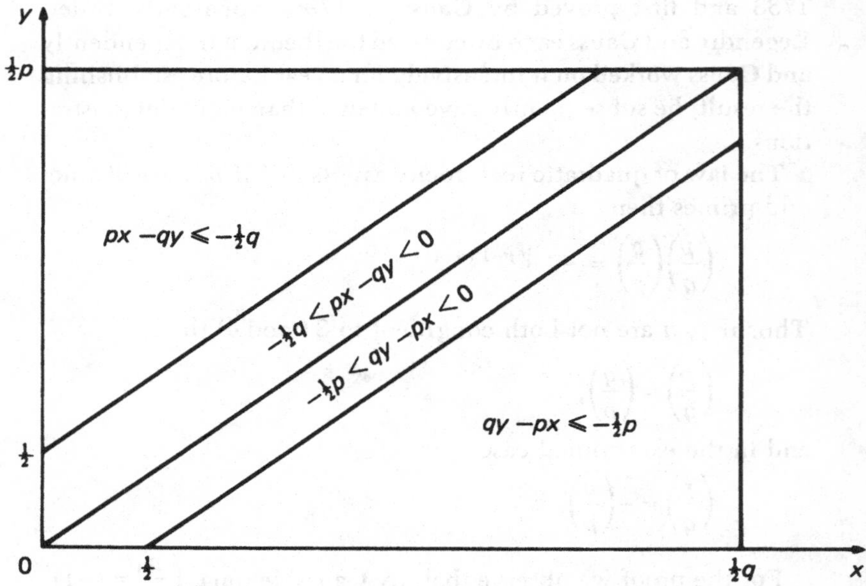

Fig. 4.1. The rectangle R in the proof of the law of quadratic reciprocity.

furnishes a one–one correspondence between them. The theorem follows.

The law of quadratic reciprocity is useful in the calculation of Legendre symbols. For example, we have

$$\left(\frac{15}{71}\right)=\left(\frac{3}{71}\right)\left(\frac{5}{71}\right)=-\left(\frac{71}{3}\right)\left(\frac{71}{5}\right)=-\left(\frac{2}{3}\right)\left(\frac{1}{5}\right)=1.$$

Further, for instance, we obtain

$$\left(\frac{-3}{p}\right)=\left(\frac{-1}{p}\right)\left(\frac{3}{p}\right)=(-1)^{\frac{1}{2}(p-1)}\left(\frac{3}{p}\right)=\left(\frac{p}{3}\right),$$

whence -3 is a quadratic residue of all primes $\equiv 1 \pmod 6$ and a quadratic non-residue of all primes $\equiv -1 \pmod 6$.

5 Jacobi's symbol

This is a generalization of the Legendre symbol. Let n be a positive odd integer and suppose that $n = p_1 p_2 \cdots p_k$ as a product of primes, not necessarily distinct. Then, for any integer a with $(a, n) = 1$, the Jacobi symbol is defined by

$$\left(\frac{a}{n}\right)=\left(\frac{a}{p_1}\right)\cdots\left(\frac{a}{p_k}\right),$$

where the factors on the right are Legendre symbols. When $n = 1$ the Jacobi symbol is defined as 1 and when $(a, n) > 1$ it is defined as 0. Clearly, if $a \equiv a' \pmod n$ then

$$\left(\frac{a}{n}\right)=\left(\frac{a'}{n}\right).$$

It should be noted at once that

$$\left(\frac{a}{n}\right)=1$$

does not imply that a is a quadratic residue (mod n). Indeed a is a quadratic residue (mod n) if and only if a is a quadratic residue (mod p) for each prime divisor p of n (see § 5 of Chapter 3). But

$$\left(\frac{a}{n}\right)=-1$$

does imply that a is a quadratic non-residue (mod n). Thus, for example, since

$$\left(\frac{6}{35}\right)=\left(\frac{6}{5}\right)\left(\frac{6}{7}\right)=\left(\frac{1}{5}\right)\left(\frac{-1}{7}\right)=-1,$$

we conclude that 6 is a quadratic non-residue (mod 35).

The Jacobi symbol is multiplicative, like the Legendre symbol; that is

$$\left(\frac{ab}{n}\right)=\left(\frac{a}{n}\right)\left(\frac{b}{n}\right)$$

for all integers a, b relatively prime to n. Further, if m, n are odd and $(a, mn)=1$ then

$$\left(\frac{a}{mn}\right)=\left(\frac{a}{m}\right)\left(\frac{a}{n}\right).$$

Furthermore we have

$$\left(\frac{-1}{n}\right)=(-1)^{\frac{1}{2}(n-1)}, \qquad \left(\frac{2}{n}\right)=(-1)^{\frac{1}{8}(n^2-1)},$$

and the analogue of the law of quadratic reciprocity holds, namely if m, n are odd and $(m, n)=1$ then

$$\left(\frac{m}{n}\right)\left(\frac{n}{m}\right)=(-1)^{\frac{1}{4}(m-1)(n-1)}.$$

These results are readily verified from the corresponding theorems for the Legendre symbol, on noting that, if $n=n_1n_2$, then

$$\tfrac{1}{2}(n-1)\equiv\tfrac{1}{2}(n_1-1)+\tfrac{1}{2}(n_2-1) \pmod 2,$$

since $\frac{1}{2}(n_1-1)(n_2-1)\equiv 0 \pmod 2$, and that a similar congruence holds for $\frac{1}{8}(n^2-1)$.

Jacobi symbols can be used to facilitate the calculation of Legendre symbols. We have, for example,

$$\left(\frac{335}{2999}\right)=-\left(\frac{2999}{335}\right)=-\left(\frac{-16}{335}\right)=-\left(\frac{-1}{335}\right)=1,$$

whence, since 2999 is a prime, it follows that 335 is a quadratic residue (mod 2999).

6 Further reading

The theories here date back to the *Disquisitiones arithmeticae* of Gauss, and they are covered by numerous texts. An excellent account of the history relating to the law of quadratic reciprocity is given by Bachmann, *Niedere Zahlentheorie* (Teubner, Leipzig, 1902), Vol. 1. In particular he gives references to some forty different proofs. For an account of modern developments associated with the law of quadratic reciprocity see Artin and Tate, *Class field theory* (W.A. Benjamin Inc., New York, 1967) and Cassels and Fröhlich (Editors) *Algebraic number theory* (Academic Press, London, 1967).

The study of higher congruences, that is congruences of the form $f(x_1, \ldots, x_n) \equiv 0 \pmod{p^j}$, where f is a polynomial with integer coefficients, leads to the concept of p-adic numbers and to deep theories in the realm of algebraic geometry; see, for example, Borevich and Shafarevich, *Number theory* (Academic Press, London, 1966), and Weil, 'Numbers of solutions of equations in finite fields', *Bull. American Math. Soc.* **55** (1949), 497–508.

7 Exercises

(i) Determine the primes p for which 5 is a quadratic residue (mod p).

(ii) Show that if p is a prime $\equiv 3 \pmod 4$ and if $p' = 2p+1$ is a prime then $2^p \equiv 1 \pmod{p'}$. Deduce that $2^{251} - 1$ is not a Mersenne prime.

(iii) Show that if p is an odd prime then the product P of all the quadratic residues (mod p) satisfies $P \equiv (-1)^{\frac{1}{2}(p+1)} \pmod p$.

(iv) Prove that if p is a prime $\equiv 1 \pmod 4$ then $\sum r = \frac{1}{4}p(p-1)$, where the summation is over all quadratic residues r with $1 \le r \le p-1$.

(v) Evaluate the Jacobi symbol $\left(\frac{123}{917}\right)$.

(vi) Show that, for any integer d and any odd prime p, the number of solutions of the congruence $x^2 \equiv d \pmod p$ is $1 + \left(\frac{d}{p}\right)$.

(vii) Let $f(x)=ax^2+bx+c$, where a, b, c are integers, and let p be an odd prime that does not divide a. Further let $d=b^2-4ac$. Show that, if p does not divide d, then

$$\sum_{x=1}^{p}\left(\frac{f(x)}{p}\right)=-\left(\frac{a}{p}\right).$$

Evaluate the sum when p divides d.

(viii) Prove that if p' is a prime $\equiv 1 \pmod 4$ and if $p=2p'+1$ is a prime then 2 is a primitive root (mod p). For which primes p' with $p=2p'+1$ prime is 5 a primitive root (mod p)?

(ix) Show that if p is a prime and a, b, c are integers not divisible by p then there are integers x, y such that $ax^2+by^2\equiv c \pmod p$.

(x) Let $f=f(x_1, \ldots, x_n)$ be a polynomial with integer coefficients that vanishes at the origin and let p be a prime. Prove that if the congruence $f\equiv 0 \pmod p$ has only the trivial solution then the polynomial

$$1-f^{p-1}-(1-x_1^{p-1})\cdots(1-x_n^{p-1})$$

is divisible by p for all integers $x_1, \ldots, x_n$. Deduce that if f has total degree less than n then the congruence $f\equiv 0 \pmod p$ has a non-trivial solution (Chevalley's theorem).

(xi) Prove that if $f=f(x_1, \ldots, x_n)$ is a quadratic form with integer coefficients, if $n\geq 3$, and if p is a prime then the congruence $f\equiv 0 \pmod p$ has a non-trivial solution.

5

Quadratic forms

1 Equivalence

We shall consider binary quadratic forms

$$f(x, y) = ax^2 + bxy + cy^2,$$

where a, b, c are integers. By the discriminant of f we mean the number $d = b^2 - 4ac$. Plainly $d \equiv 0 \pmod 4$ if b is even and $d \equiv 1 \pmod 4$ if b is odd. The forms $x^2 - \frac{1}{4}dy^2$ for $d \equiv 0 \pmod 4$ and $x^2 + xy + \frac{1}{4}(1-d)y^2$ for $d \equiv 1 \pmod 4$ are called the principal forms with discriminant d. We have

$$4af(x, y) = (2ax + by)^2 - dy^2,$$

whence if $d < 0$ the values taken by f are all of the same sign (or zero); f is called positive or negative definite accordingly. If $d > 0$ then f takes values of both signs and it is called indefinite.

We say that two quadratic forms are equivalent if one can be transformed into the other by an integral unimodular substitution, that is, a substitution of the form

$$x = px' + qy', \qquad y = rx' + sy',$$

where p, q, r, s are integers with $ps - qr = 1$. It is readily verified that this relation is reflexive, symmetric and transitive. Further, it is clear that the set of values assumed by equivalent forms as x, y run through the integers are the same, and indeed they assume the same set of values as the pair x, y runs through all relatively prime integers; for $(x, y) = 1$ if and only if $(x', y') = 1$. Furthermore equivalent forms have the same discriminant. For the substitution takes f into

$$f'(x', y') = a'x'^2 + b'x'y' + c'y'^2,$$

where

$$a' = f(p, r), \quad b' = 2apq + b(ps + qr) + 2crs,$$
$$c' = f(q, s),$$

and it is readily checked that $b'^2 - 4a'c' = d(ps - qr)^2$. Alternatively, in matrix notation, we can write f as X^TFX and the substitution as $X = UX'$, where

$$X = \begin{pmatrix} x \\ y \end{pmatrix}, \quad X' = \begin{pmatrix} x' \\ y' \end{pmatrix}, \quad F = \begin{pmatrix} a & \frac{1}{2}b \\ \frac{1}{2}b & c \end{pmatrix}, \quad U = \begin{pmatrix} p & q \\ r & s \end{pmatrix};$$

then f is transformed into $X'^TF'X'$, where $F' = U^TFU$, and, since the determinant of U is 1, it follows that the determinants of F and F' are equal.

2 Reduction

There is an elegant theory of reduction relating to positive definite quadratic forms which we shall now describe. Accordingly we shall assume henceforth that $d < 0$ and that $a > 0$; then we have also $c > 0$.

We begin by observing that by a finite sequence of unimodular substitutions of the form $x = y'$, $y = -x'$ and $x = x' \pm y'$, $y = y'$, f can be transformed into another binary form for which $|b| \leq a \leq c$. For the first of these substitutions interchanges a and c whence it allows one to replace $a > c$ by $a < c$; and the second has the effect of changing b to $b \pm 2a$, leaving a unchanged, whence, by finitely many applications it allows one to replace $|b| > a$ by $|b| \leq a$. The process must terminate since, whenever the first substitution is applied it results in a smaller value of a. In fact we can transform f into a binary form for which either

$$-a < b \leq a < c \quad \text{or} \quad 0 \leq b \leq a = c.$$

For if $b = -a$ then the second of the above substitutions allows one to take $b = a$, leaving c unchanged, and if $a = c$ then the first substitution allows one to take $0 \leq b$. A binary form for which one or other of the above conditions on a, b, c holds is said to be reduced.

There are only finitely many reduced forms with a given discriminant d; for if f is reduced then $-d = 4ac - b^2 \geq 3ac$, whence a, c and $|b|$ cannot exceed $\frac{1}{3}|d|$. The number of reduced forms with discriminant d is called the class number and it is denoted by $h(d)$. To calculate the class number when $d=-4$, for example, we note that the inequality $3ac \le 4$ gives $a=c=1$, whence $b=0$ and $h(-4)=1$. The number $h(d)$ is actually the number of inequivalent classes of binary quadratic forms with discriminant d since, as we shall now prove, any two reduced forms are not equivalent.

Let $f(x, y)$ be a reduced form. Then if x, y are non-zero integers and $|x| \ge |y|$ we have

$$\begin{aligned} f(x, y) &\ge |x|(a|x| - |by|) + c|y|^2 \\ &\ge |x|^2(a-|b|) + c|y|^2 \ge a - |b| + c. \end{aligned}$$

Similarly if $|y| \ge |x|$ we have $f(x, y) \ge a-|b|+c$. Hence the smallest values assumed by f for relatively prime integers x, y are a, c and $a-|b|+c$ in that order; these values are taken at $(1, 0)$, $(0, 1)$ and either $(1, 1)$ or $(1, -1)$. Now the sequences of values assumed by equivalent forms for relatively prime x, y are the same, except for a rearrangement, and thus if f' is a form, as in § 1, equivalent to f, and if also f' is reduced, then $a=a'$, $c=c'$ and $b=\pm b'$. It remains therefore to prove that if $b=-b'$ then in fact $b=0$. We can assume here that $-a<b<a<c$, for, since f' is reduced, we have $-a<-b$, and if $a=c$ then we have $b \ge 0$, $-b \ge 0$, whence $b=0$. It follows that $f(x, y) \ge a-|b|+c>c>a$ for all non-zero integers x, y. But, with the notation of § 1 for the substitution taking f to f', we have $a=f(p, r)$. Thus $p=\pm 1$, $r=0$, and from $ps-qr=1$ we obtain $s=\pm 1$. Further we have $c=f(q, s)$ whence $q=0$. Hence the only substitutions taking f to f' are $x=x'$, $y=y'$ and $x=-x'$, $y=-y'$. These give $b=0$, as required.

3 Representations by binary forms

A number n is said to be properly represented by a binary form f if $n=f(x, y)$ for some integers x, y with $(x, y)=1$. There is a useful criterion in connection with such representations, namely n is properly represented by some binary form with discriminant d if and only if the congruence $x^2 \equiv d \pmod{4n}$ is soluble.

For the proof, suppose first that the congruence is soluble and let $x=b$ be a solution. Define c by $b^2-4nc=d$ and put $a=n$. Then the form f, as in § 1, has discriminant d and it properly represents n, in fact $f(1,0)=n$. Conversely suppose that f has discriminant d and that $n=f(p,r)$ for some integers p, r with $(p,r)=1$. Then there exist integers r, s with $ps-qr=1$ and f is equivalent to a form f' as in § 1 with $a'=n$. But f and f' have the same discriminants and so $b'^2-4nc'=d$. Hence the congruence $x^2\equiv d \pmod{4n}$ has a solution $x=b'$.

The ideas here can be developed to furnish, in the case $(n,d)=1$, the number of proper representations of n by all reduced forms with a given discriminant d. Indeed the quantity in question is given by ws, where s is the number of solutions of the congruence $x^2\equiv d \pmod{4n}$ with $0\leq x<2n$ and w is the number of automorphs of a reduced form; by an automorph of f we mean an integral unimodular substitution that takes f into itself. The number w is related to the solutions of the Pell equation (see § 3 of Chapter 7); it is given by 2 for $d<-4$, by 4 for $d=-4$ and by 6 for $d=-3$. In fact the only automorphs, for $d<-4$, are $x=x'$, $y=y'$ and $x=-x'$, $y=-y'$.

4 Sums of two squares

Let n be a natural number. We proceed to prove that n can be expressed in the form x^2+y^2 for some integers x, y if and only if every prime divisor p of n with $p\equiv 3 \pmod 4$ occurs to an even power in the standard factorization of n. The result dates back to Fermat and Euler.

The necessity is easily verified, for suppose that $n=x^2+y^2$ and that n is divisible by a prime $p\equiv 3 \pmod 4$. Then $x^2\equiv -y^2 \pmod p$ and since -1 is a quadratic non-residue $(\text{mod } p)$, we see that p divides x and y. Thus we have $(x/p)^2+(y/p)^2=n/p^2$, and it follows by induction that p divides n to an even power.

To prove the converse it will suffice to assume that n is square free and to show that if each odd prime divisor p of n satisfies $p\equiv 1 \pmod 4$ then n can be represented by x^2+y^2; for clearly if $n=x^2+y^2$ then $nm^2=(xm)^2+(ym)^2$. Now the quadratic form x^2+y^2 is a reduced form with discriminant -4, and it was proved in § 2 that $h(-4)=1$. Hence it is the only such reduced form. It follows from § 3 that n is properly represented by x^2+y^2 if and only if the congruence $x^2\equiv -4 \pmod{4n}$ is soluble. But, by hypothesis, -1 is a quadratic residue (mod p) for each prime divisor p of n. Hence -1 is a quadratic residue (mod n) and the result follows.

It will be noted that the argument involves the Chinese remainder theorem; but this can be avoided by appeal to the identity

$$(x^2+y^2)(x'^2+y'^2)=(xx'+yy')^2+(xy'-yx')^2$$

which enables one to consider only prime values of n. In fact there is a well known proof of the theorem based on this identity alone, similar to § 5 below.

The demonstration here can be refined to furnish the number of representations of n as x^2+y^2. The number is given by $4\sum\left(\frac{-1}{m}\right)$, where the summation is over all odd divisors m of n. Thus, for instance, each prime $p\equiv 1 \pmod 4$ can be expressed in precisely eight ways as the sum of two squares.

5 Sums of four squares

We prove now the famous theorem stated by Bachet in 1621 and first demonstrated by Lagrange in 1770 to the effect that every natural number can be expressed as the sum of four integer squares. Our proof will be based on the identity

$$\begin{aligned}(x^2+y^2+z^2+w^2)&(x'^2+y'^2+z'^2+w'^2)\\ &=(xx'+yy'+zz'+ww')^2+(xy'-yx'+wz'-zw')^2\\ &\quad+(xz'-zx'+yw'-wy')^2+(xw'-wx'+zy'-yz')^2,\end{aligned}$$

which is related to the theory of quaternions.

In view of the identity and the trivial representation $2=1^2+1^2+0^2+0^2$, it will suffice to prove the theorem for odd primes p. Now the numbers x^2 with $0\le x\le\frac{1}{2}(p-1)$ are mutually incongruent (mod p), and the same holds for the numbers $-1-y^2$ with $0\le y\le\frac{1}{2}(p-1)$. Thus we have $x^2\equiv -1-y^2 \pmod p$ for some x, y satisfying $x^2+y^2+1<1+2(\frac{1}{2}p)^2<p^2$. Hence we obtain $mp=x^2+y^2+1$ for some integer m with $0<m<p$.

Let l be the least positive integer such that $lp=x^2+y^2+z^2+w^2$ for some integers x, y, z, w. Then $l\le m<p$. Further l is odd, for if l were even then an even number of x, y, z, w would be odd

and we could assume that $x+y$, $x-y$, $z+w$, $z-w$ are even; but

$$\tfrac{1}{2}lp=(\tfrac{1}{2}(x+y))^2+(\tfrac{1}{2}(x-y))^2+(\tfrac{1}{2}(z+w))^2+(\tfrac{1}{2}(z-w))^2$$

and this is inconsistent with the minimal choice of l. To prove the theorem we have to show that $l=1$; accordingly we suppose that $l>1$ and obtain a contradiction. Let x', y', z', w' be the numerically least residues of x, y, z, w (mod l) and put

$$n=x'^2+y'^2+z'^2+w'^2.$$

Then $n\equiv 0 \pmod{l}$ and we have $n>0$, for otherwise l would divide p. Further, since l is odd, we have $n<4(\tfrac{1}{2}l)^2=l^2$. Thus $n=kl$ for some integer k with $0<k<l$. Now by the identity we see that $(kl)(lp)$ is expressible as a sum of four integer squares, and moreover it is clear that each of these squares is divisible by l^2. Thus kp is expressible as a sum of four integer squares. But this contradicts the definition of l and the theorem follows. The argument here is an illustration of Fermat's method of infinite descent.

There is a result dating back to Legendre and Gauss to the effect that a natural number is the sum of three squares if and only if it is not of the form $4^j(8k+7)$ with j, k non-negative integers. Here the necessity is obvious since a square is congruent to 0, 1 or 4 (mod 8) but the sufficiency depends on the theory of ternary quadratic forms.

Waring conjectured in 1770 that every natural number can be represented as the sum of four squares, nine cubes, nineteen biquadrates 'and so on'. One interprets the latter to mean that, for every integer $k\geq 2$ there exists an integer $s=s(k)$ such that every natural number n can be expressed in the form $x_1{}^k+\cdots+x_s{}^k$ with $x_1,\ldots,x_s$ non-negative integers; and it is customary to denote the least such s by $g(k)$. Thus we have $g(2)=4$. Waring's conjecture was proved by Hilbert in 1909. Another, quite different proof was given by Hardy and Littlewood in 1920 and it was here that they described for the first time their famous 'circle method'. The work depends on the identity

$$\sum_{n=0}^{\infty} r(n)z^n=(f(z))^s,$$

where $r(n)$ denotes the number of representations of n in the required form and $f(z)=1+z^{1^k}+z^{2^k}+\cdots$. Thus we have

$$r(n)=\frac{1}{2\pi i}\int_C \frac{(f(z))^s}{z^{n+1}}\,dz$$

for a suitable contour C. The argument now involves a delicate division of the contour into 'major and minor' arcs, and the analysis leads to an asymptotic expression for $r(n)$ and to precise estimates for $g(k)$.

6 Further reading

A careful account of the theory of binary quadratic forms is given in Landau, *Elementary number theory* (Chelsea Publ. Co., New York, 1966); see also Davenport, *The higher arithmetic* (Cambridge U.P., 5th edn, 1982). As there, we have used the classical definition of equivalence in terms of substitutions with determinant 1; however, there is an analogous theory involving substitutions with determinant ±1 and this is described in Niven and Zuckerman, *An introduction to the theory of numbers* (Wiley, New York, 4th edn, 1980).

For a comprehensive account of the general theory of quadratic forms see Cassels, *Rational quadratic forms* (Academic Press, London and New York, 1978). For an account of the analysis appertaining to Waring's problem see R. C. Vaughan, *The Hardy–Littlewood method* (Cambridge U.P., 1981).

7 Exercises

(i) Prove that $h(d)=1$ when $d=-3, -4, -7, -8, -11, -19, -43, -67$ and -163.

(ii) Determine all the odd primes that can be expressed in the form $x^2+xy+5y^2$.

(iii) Determine all the positive integers that can be expressed in the form x^2+2y^2.

(iv) Determine all the positive integers that can be expressed in the form x^2-y^2.

(v) Show that there are precisely two reduced forms with discriminant -20. Hence prove that the primes that can be represented by x^2+5y^2 are 5 and those congruent to 1 or 9 (mod 20).

(vi) Calculate $h(-31)$.

(vii) Find the least positive integer that can be represented by $4x^2+17xy+20y^2$.

(viii) Prove that n and $2n$, where n is any positive integer, have the same number of representations as the sum of two squares.

(ix) Find the least integer s such that $n=x_1{}^k+\cdots+x_s{}^k$, where $n=2^k[(3/2)^k]-1$ and $x_1,\ldots,x_s$ are positive integers.

6

Diophantine approximation

1 Dirichlet's theorem

Diophantine approximation is concerned with the solubility of inequalities in integers. The simplest result in this field was obtained by Dirichlet in 1842. He showed that, for any real θ and any integer $Q>1$ there exist integers p, q with $0<q<Q$ such that $|q\theta-p|\le 1/Q$.

The result can be derived at once from the so-called 'box' or 'pigeon-hole' principle. This asserts that if there are n holes containing $n+1$ pigeons then there must be at least two pigeons in some hole. Consider in fact the $Q+1$ numbers 0, 1, $\{\theta\}$, $\{2\theta\},\ldots,\{(Q-1)\theta\}$, where $\{x\}$ denotes the fractional part of x as in Chapter 2. These numbers all lie in the interval [0, 1], and if one divides the latter, as clearly one can, into Q disjoint sub-intervals, each of length $1/Q$, then it follows that two of the $Q+1$ numbers must lie in one of the Q sub-intervals. The difference between the two numbers has the form $q\theta-p$, where p, q are integers with $0<q<Q$, and we have $|q\theta-p|\le 1/Q$, as required.

Dirichlet's theorem holds more generally for any real $Q>1$; the result for non-integral Q follows from the theorem just established with Q replaced by $[Q]+1$. Further it is clear that the integers p, q referred to in the theorem can be chosen to be relatively prime. When θ is irrational we have the important corollary that there exist infinitely many rationals p/q $(q>0)$ such that $|\theta-p/q|<1/q^2$. Indeed, for $Q>1$, there is a rational p/q with $|\theta-p/q|\le 1/(Qq)<1/q^2$; moreover, if θ is irrational then, for any Q' exceeding $1/|q\theta-p|$, the rational corresponding to Q' will be different from p/q. Note that the corollary does not remain valid for rational θ; for if $\theta=a/b$ with a, b integers and $b>0$ then, when $\theta \neq p/q$, we have $|\theta - p/q| \geq 1/(qb)$ and so there are only finitely many rationals p/q such that $|\theta - p/q| < 1/q^2$.

2 Continued fractions

The continued fraction algorithm sets up a one–one correspondence between all irrational θ and all infinite sets of integers $a_0, a_1, a_2, \ldots$ with $a_1, a_2, \ldots$ positive. It also sets up a one–one correspondence between all rational θ and all finite sets of integers $a_0, a_1, \ldots, a_n$ with $a_1, a_2, \ldots, a_{n-1}$ positive and with $a_n \geq 2$.

To describe the algorithm, let θ be any real number. We put $a_0 = [\theta]$. If $a_0 \neq \theta$ we write $\theta = a_0 + 1/\theta_1$, so that $\theta_1 > 1$, and we put $a_1 = [\theta_1]$. If $a_1 \neq \theta_1$ we write $\theta_1 = a_1 + 1/\theta_2$, so that $\theta_2 > 1$, and we put $a_2 = [\theta_2]$. The process continues indefinitely unless $a_n = \theta_n$ for some n. It is clear that if the latter occurs then θ is rational; in fact we have

$$\theta = a_0 + \cfrac{1}{a_1 + \cfrac{1}{a_2 + \genfrac{}{}{0pt}{}{}{\ddots \; \cfrac{1}{a_n}}}}.$$

Conversely, as will be clear in a moment, if θ is rational then the process terminates. The expression above is called the continued fraction for θ; it is customary to write the equation briefly as

$$\theta = a_0 + \frac{1}{a_1+}\,\frac{1}{a_2+}\cdots\frac{1}{a_n}$$

or, more briefly, as

$$\theta = [a_0, a_1, a_2, \ldots, a_n].$$

If $a_n \neq \theta_n$ for all n, so that the process does not terminate, then θ is irrational. We proceed to show that one can then write

$$\theta = a_0 + \frac{1}{a_1+}\,\frac{1}{a_2+}\cdots,$$

or briefly

$$\theta = [a_0, a_1, a_2, \ldots].$$

The integers $a_0, a_1, a_2, \ldots$ are known as the partial quotients of θ; the numbers $\theta_1, \theta_2, \ldots$ are referred to as the complete quotients of θ. We shall prove that the rationals

$$p_n/q_n = [a_0, a_1, \ldots, a_n],$$

where p_n, q_n denote relatively prime integers, tend to θ as $n \to \infty$; they are in fact known as the convergents to θ.

First we show that the p_n, q_n are generated recursively by the equations

$$p_n = a_n p_{n-1} + p_{n-2}, \qquad q_n = a_n q_{n-1} + q_{n-2},$$

where $p_0 = a_0$, $q_0 = 1$ and $p_1 = a_0 a_1 + 1$, $q_1 = a_1$. The recurrences plainly hold for $n = 2$; we assume that they hold for $n = m - 1 \geq 2$ and we proceed to verify them for $n = m$. We define relatively prime integers p'_j, q'_j $(j = 0, 1, \ldots)$ by

$$p'_j/q'_j = [a_1, a_2, \ldots, a_{j+1}],$$

and we apply the recurrences to p'_{m-1}, q'_{m-1}; they give

$$p'_{m-1} = a_m p'_{m-2} + p'_{m-3}, \quad q'_{m-1} = a_m q'_{m-2} + q'_{m-3}.$$

But we have $p_j/q_j = a_0 + q'_{j-1}/p'_{j-1}$, whence

$$p_j = a_0 p'_{j-1} + q'_{j-1}, \quad q_j = p'_{j-1}.$$

Thus, on taking $j = m$, we obtain

$$p_m = a_m(a_0 p'_{m-2} + q'_{m-2}) + a_0 p'_{m-3} + q'_{m-3},$$

$$q_m = a_m p'_{m-2} + p'_{m-3},$$

and, on taking $j = m - 1$ and $j = m - 2$, it follows that

$$p_m = a_m p_{m-1} + p_{m-2}, \qquad q_m = a_m q_{m-1} + q_{m-2},$$

as required.

Now by the definition of $\theta_1, \theta_2, \ldots$ we have

$$\theta = [a_0, a_1, \ldots, a_n, \theta_{n+1}],$$

where $0 < 1/\theta_{n+1} \leq 1/a_{n+1}$; hence θ lies between p_n/q_n and p_{n+1}/q_{n+1}. It is readily seen by induction that the above recurrences give

$$p_n q_{n+1} - p_{n+1} q_n = (-1)^{n+1},$$

and thus we have

$$|p_n/q_n - p_{n+1}/q_{n+1}| = 1/(q_n q_{n+1}).$$

It follows that the convergents p_n/q_n to θ satisfy

$$|\theta - p_n/q_n| \leq 1/(q_n q_{n+1}),$$

and so certainly $p_n/q_n \to \theta$ as $n \to \infty$.

In view of the latter inequality and the remarks at the end of § 1, it is now clear that when θ is rational the continued fraction process terminates. Indeed, for rational θ, the process is closely related to Euclid's algorithm as described in Chapter 1. In fact, if, with the notation of § 4 of Chapter 1, we take $\theta = a/b$ and $a_j = q_{j+1}$ $(0 \leq j \leq k)$ then we have

$$\theta = [a_0, a_1, \ldots, a_k];$$

thus, for example, $\frac{187}{35} = [5, 2, 1, 11]$.

3 Rational approximations

It follows from the results of § 2 that, for any real θ, each convergent p/q satisfies $|\theta - p/q| < 1/q^2$. We observe now that, of any two consecutive convergents, say p_n/q_n and p_{n+1}/q_{n+1}, one at least satisfies $|\theta - p/q| < 1/(2q^2)$. Indeed, since $\theta - p_n/q_n$ and $\theta - p_{n+1}/q_{n+1}$ have opposite signs, we have

$$|\theta - p_n/q_n| + |\theta - p_{n+1}/q_{n+1}| = |p_n/q_n - p_{n+1}/q_{n+1}|$$
$$= 1/(q_n q_{n+1});$$

but, for any real α, β with $\alpha \neq \beta$, we have $\alpha\beta < \frac{1}{2}(\alpha^2 + \beta^2)$, whence

$$1/(q_n q_{n+1}) < 1/(2q_n^2) + 1/(2q_{n+1}^2),$$

and this gives the required result. We observe further that, of any three consecutive convergents, say p_n/q_n, p_{n+1}/q_{n+1} and p_{n+2}/q_{n+2}, one at least satisfies $|\theta - p/q| < 1/(\sqrt{5}\, q^2)$. In fact, if the result were false, then the equations above would give

$$1/(\sqrt{5}\, q_n^2) + 1/(\sqrt{5}\, q_{n+1}^2) \leq 1/(q_n q_{n+1}),$$

that is $\lambda + 1/\lambda \leq \sqrt{5}$, where $\lambda = q_{n+1}/q_n$. Since λ is rational it follows that strict inequality holds and so

$$(\lambda - \tfrac{1}{2}(1 + \sqrt{5}))(\lambda + \tfrac{1}{2}(1 - \sqrt{5})) < 0,$$

whence $\lambda < \frac{1}{2}(1 + \sqrt{5})$. Similarly, on writing $\mu = q_{n+2}/q_{n+1}$, we would have $\mu < \frac{1}{2}(1 + \sqrt{5})$. But, by § 2, we have $q_{n+2} = a_{n+2}q_{n+1} + q_n$, and thus $\mu \geq 1 + 1/\lambda$; this gives a contradiction, for if $\lambda < \frac{1}{2}(1 + \sqrt{5})$ then $1/\lambda > \frac{1}{2}(-1 + \sqrt{5})$.

The latter result confirms a theorem of Hurwitz to the effect that, for any irrational θ, there exist infinitely many rational p/q such that $|\theta - p/q| < 1/(\sqrt{5}\, q^2)$. The constant $1/\sqrt{5}$ is best possible, as can be verified (see § 5) by taking

$$\theta = \tfrac{1}{2}(1 + \sqrt{5}) = [1, 1, 1, \ldots].$$

However if one excludes all irrationals equivalent to θ, that is those whose continued fractions have all but finitely many partial quotients equal to 1, then Hurwitz's theorem holds with $1/\sqrt{8}$ in place of $1/\sqrt{5}$, and this is again best possible. There is an infinite sequence of such results, with constants tending to 1/3, and they constitute the so-called Markoff chain.

We note next that the convergents give successively closer approximations to θ. In fact we have the stronger result that $|q_n\theta - p_n|$ decreases as n increases. To verify this, we observe that the recurrences in § 2 hold for any indeterminates $a_0, a_1, \ldots$, whence, for $n \geq 1$, we have

$$\theta = \frac{p_n\theta_{n+1} + p_{n-1}}{q_n\theta_{n+1} + q_{n-1}};$$

thus we obtain

$$|q_n\theta - p_n| = 1/(q_n\theta_{n+1} + q_{n-1}),$$

and the assertion follows since, for $n > 1$, the denominator on the right exceeds

$$q_n + q_{n-1} = (a_n + 1)q_{n-1} + q_{n-2} > q_{n-1}\theta_n + q_{n-2},$$

and, for $n = 1$, it exceeds θ_1. The argument here shows, incidentally, that the convergents to θ satisfy

$$\frac{1}{(a_{n+1}+2)q_n^2} < \left|\theta - \frac{p_n}{q_n}\right| < \frac{1}{a_{n+1}q_n^2}.$$

The convergents are indeed the best approximations to θ in the sense that, if p, q are integers with $0 < q < q_{n+1}$ then $|q\theta - p| \geq |q_n\theta - p_n|$. For if we define integers u, v by

$$p = up_n + vp_{n+1}, \qquad q = uq_n + vq_{n+1},$$

then it is easily seen that $u \neq 0$ and that, if $v \neq 0$, then u, v have opposite signs; hence, since $q_n\theta - p_n$ and $q_{n+1}\theta - p_{n+1}$ have opposite signs, we obtain

$$\begin{aligned} |q\theta - p| &= |u(q_n\theta - p_n) + v(q_{n+1}\theta - p_{n+1})| \\ &\geq |q_n\theta - p_n|, \end{aligned}$$

as required. As a corollary, we deduce that if a rational p/q satisfies $|\theta - p/q| < 1/(2q^2)$ then it is a convergent to θ. In fact we have $p/q = p_n/q_n$, where $q_n \leq q < q_{n+1}$; for clearly

$$\begin{aligned} |p/q - p_n/q_n| &\leq |\theta - p/q| + |\theta - p_n/q_n| \\ &\leq (1/q + 1/q_n)|q\theta - p|, \end{aligned}$$

and, since $q \geq q_n$ and $|q\theta - p| < 1/(2q)$, the number on the right is less than $1/(qq_n)$; hence the number on the left vanishes, as required.

To conclude this section we remark that, for almost all real θ in the sense of Lebesgue measure, the inequality $|\theta - p/q| < 1/(q^2 \log q)$ has infinitely many rational solutions p/q; in fact the same applies to the inequality $|\theta - p/q| < f(q)/q$, where f is any monotonically decreasing function such that $\sum f(q)$ diverges. However, almost no θ have the property if $\sum f(q)$ converges, for instance if $f(q) = 1/(q(\log q)^{1+\delta})$ with $\delta > 0$.

4 Quadratic irrationals

By a quadratic irrational we mean a zero of a polynomial $ax^2 + bx + c$, where a, b, c are integers and the discriminant $d = b^2 - 4ac$ is positive and not a perfect square. One of the most remarkable results in the theory of numbers, known since the time of Lagrange, is that a continued fraction represents a quadratic irrational if and only if it is ultimately periodic, that is, if and only if the partial quotients $a_0, a_1, \ldots$ satisfy $a_{m+n} = a_n$ for some positive integer m and for all sufficiently large n. Thus a continued fraction θ is a quadratic irrational if and only if it has the form

$$\theta = [a_0, a_1, \ldots, a_{k-1}, \overline{a_k, \ldots, a_{k+m-1}}],$$

where the bar indicates that the block of partial quotients is repeated indefinitely. As examples, we have $\sqrt{2} = [1, \bar{2}]$ and $\frac{1}{3}(3 + \sqrt{3}) = [1, 1, \overline{1, 2}]$.

It is easy to see that if the continued fraction for θ has the above form then θ is a quadratic irrational. For the number

$$\phi = [\overline{a_k, \ldots, a_{k+m-1}}]$$

is a complete quotient of θ and so, by § 3, we have, for $k \geq 2$,

$$\theta = \frac{p_{k-1}\phi + p_{k-2}}{q_{k-1}\phi + q_{k-2}},$$

where p_n/q_n $(n = 0, 1, \ldots)$ are the convergents to θ; further we have, for $m \geq 2$,

$$\phi = \frac{p'_{m-1}\phi + p'_{m-2}}{q'_{m-1}\phi + q'_{m-2}},$$

where p'_n/q'_n $(n=0, 1, \ldots)$ are the convergents to ϕ. It is clear from the latter equation that ϕ is quadratic and hence, by the preceding equation, so also is θ; and this plainly remains valid for $k=0$ and 1, and for $m=1$. Since the continued fraction for θ does not terminate, it follows that θ is a quadratic irrational, as required.

To prove the converse, suppose that θ is a quadratic irrational so that θ satisfies an equation $ax^2+bx+c=0$, where a, b, c are integers with $d=b^2-4ac>0$. We shall consider the binary form

$$f(x, y)=ax^2+bxy+cy^2.$$

The substitution

$$x=p_nx'+p_{n-1}y', \qquad y=q_nx'+q_{n-1}y',$$

where p_n/q_n $(n=1, 2, \ldots)$ denote the convergents to θ, has determinant

$$p_nq_{n-1}-p_{n-1}q_n=(-1)^{n-1},$$

and so, as in § 1 of Chapter 5, we see that it takes f into a binary form

$$f_n(x, y)=a_nx^2+b_nxy+c_ny^2$$

with the same discriminant d as f. Further we have $a_n=f(p_n, q_n)$ and $c_n=a_{n-1}$. Now $f(\theta, 1)=0$ and so

$$\begin{aligned} a_n/q_n^2 &= f(p_n/q_n, 1)-f(\theta, 1) \\ &= a((p_n/q_n)^2-\theta^2)+b((p_n/q_n)-\theta). \end{aligned}$$

By § 2 we have $|\theta-p_n/q_n|<1/q_n^2$, whence

$$|\theta^2-(p_n/q_n)^2|<|\theta+p_n/q_n|/q_n^2<(2|\theta|+1)/q_n^2.$$

Thus we see that

$$|a_n|<(2|\theta|+1)|a|+|b|,$$

that is, a_n is bounded independently of n. Since $c_n=a_{n-1}$ and $b_n^2-4a_nc_n=d$, it follows that b_n and c_n are likewise bounded. But, for $n\geq 1$, we have

$$\theta=\frac{p_n\theta_{n+1}+p_{n-1}}{q_n\theta_{n+1}+q_{n-1}},$$

where $\theta_1, \theta_2, \ldots$ denote the complete quotients of θ, and so $f_n(\theta_{n+1}, 1)=0$. This implies that there are only finitely many possibilities for $\theta_1, \theta_2, \ldots$, whence $\theta_{l+m}=\theta_l$ for some positive l, m. Hence the continued fraction for θ is ultimately periodic, as required.

The continued fraction of a quadratic irrational θ is said to be purely periodic if $k=0$ in the expression indicated above. It is easy to show that this occurs if and only if $\theta>1$ and the conjugate θ' of θ, that is, the other root of the quadratic equation defining θ, satisfies $-1<\theta'<0$. Indeed if $\theta>1$ and $-1<\theta'<0$ then it is readily verified by induction that the conjugates θ_n' of the complete quotients θ_n $(n=1,2,\ldots)$ of θ likewise satisfy $-1<\theta_n'<0$; one needs to refer only to the relation $\theta_n'=a_n+1/\theta_{n+1}'$, where $\theta=[a_0,a_1,\ldots]$, together with the fact that $a_n\geq 1$ for all n including $n=0$. The inequality $-1<\theta_n'<0$ shows that $a_n=[-1/\theta_{n+1}']$. Now since θ is a quadratic irrational we have $\theta_m=\theta_n$ for some distinct m, n; but this gives $1/\theta_m'=1/\theta_n'$, whence $a_{m-1}=a_{n-1}$. It follows that $\theta_{m-1}=\theta_{n-1}$, and repetition of this conclusion yields $\theta=\theta_{n-m}$, assuming that $n>m$. Hence θ is purely periodic. Conversely, if θ is purely periodic, then $\theta>a_0\geq 1$. Further, for some $n\geq 1$, we have

$$\theta=\frac{p_n\theta+p_{n-1}}{q_n\theta+q_{n-1}},$$

where p_n/q_n $(n=1,2,\ldots)$ denote the convergents to θ, and thus θ satisfies the equation

$$q_nx^2+(q_{n-1}-p_n)x-p_{n-1}=0.$$

Now the quadratic on the left has the value $-p_{n-1}<0$ for $x=0$, and it has the value $p_n+q_n-(p_{n-1}+q_{n-1})>0$ for $x=-1$. Hence the conjugate θ' of θ satisfies $-1<\theta'<0$, as required.

As an immediate corollary we see that the continued fractions of $\sqrt{d}+[\sqrt{d}]$ and $1/(\sqrt{d}-[\sqrt{d}])$ are purely periodic, where d is any positive integer, not a perfect square. Moreover this implies that the continued fraction of $\sqrt{d}$ is almost purely periodic in the sense that, here, $k=1$. The convergents to $\sqrt{d}$, incidentally, are closely related to the solutions of the Pell equation about which we shall speak in Chapter 8.

5 Liouville's theorem

The work of § 4 shows that every quadratic irrational θ has bounded partial quotients. It follows from the results of § 3 that there exists a number $c=c(\theta)>0$ such that the inequality $|\theta-p/q|>c/q^2$ holds for all rationals p/q $(q>0)$. Liouville proved in 1844 that a theorem of the latter kind is valid more generally for any algebraic irrational, and his discovery led to the first demonstration of the existence of transcendental numbers.

A real or complex number is said to be algebraic if it is a zero of a polynomial

$$P(x)=a_0x^n+a_1x^{n-1}+\cdots+a_n,$$

where $a_0, a_1, \ldots, a_n$ denote integers, not all 0. For each algebraic number α there is a polynomial P as above, with least degree, such that $P(\alpha)=0$, and P is unique if one assumes that $a_0>0$ and that $a_0, a_1, \ldots, a_n$ are relatively prime; obviously P is irreducible over the rationals, and it is called the minimal polynomial for α. The degree of α is defined as the degree of P.

Liouville's theorem states that for any algebraic number α with degree $n>1$ there exists a number $c=c(\alpha)>0$ such that the inequality $|\alpha-p/q|>c/q^n$ holds for all rationals p/q $(q>0)$. For the proof, we shall assume, as clearly we may, that α is real, and we shall apply the mean-value theorem to P, the minimal polynomial for α. We have, for any rational p/q $(q>0)$,

$$P(\alpha)-P(p/q)=(\alpha-p/q)P'(\xi),$$

where $P'(x)$ denotes the derivative of P, and ξ lies between α and p/q. Now we have $P(\alpha)=0$ and, since P is irreducible, we have also $P(p/q)\neq 0$. But $q^nP(p/q)$ is an integer and so $|P(p/q)|\geq 1/q^n$. We can suppose that $|\alpha-p/q|<1$, for otherwise the theorem certainly holds; then we have $|\xi|<|\alpha|+1$ and so $|P'(\xi)|<C$ for some $C=C(\alpha)$. This gives $|\alpha-p/q|>c/q^n$, where $c=1/C$, as required.

The proof here enables one to furnish an explicit value for c in terms of the degree of P and its coefficients. Let us use this observation to confirm the assertion made in § 3 concerning $\alpha=\frac{1}{2}(1+\sqrt{5})$. In this case we have $P(x)=x^2-x-1$ and so $P'(x)=2x-1$. Let p/q $(q>0)$ be any rational and let $\delta=|\alpha-p/q|$. Then $|P(p/q)|\leq\delta|P'(\xi)|$ for some ξ between α and p/q. Now clearly $|\xi|\leq\alpha+\delta$ and so

$$|P'(\xi)|\leq 2(\alpha+\delta)-1=2\delta+\sqrt{5}.$$

But $|P(p/q)|\geq 1/q^2$, whence $\delta(2\delta+\sqrt{5})\geq 1/q^2$. This implies that for any c' with $c'<1/\sqrt{5}$ and for all sufficiently large q we have $\delta>c'/q^2$. Hence Hurwitz's theorem (see § 3) is best possible.

A real or complex number that is not algebraic is said to be transcendental. It is now easy to give an example; consider, in fact, the series

$$\theta = 2^{-1!} + 2^{-2!} + 2^{-3!} + \cdots.$$

If we put

$$p_j = 2^{j!}(2^{-1!} + 2^{-2!} + \cdots + 2^{-j!}),$$
$$q_j = 2^{j!} \qquad (j = 1, 2, \ldots),$$

then p_j, q_j are integers, and we have

$$|\theta - p_j/q_j| = 2^{-(j+1)!} + 2^{-(j+2)!} + \cdots.$$

But the sum on the right is at most

$$2^{-(j+1)!}(1 + 2^{-1} + 2^{-2} + \cdots) = 2^{-(j+1)!+1} < q_j^{-j},$$

and it follows readily from Liouville's theorem that θ is transcendental. Indeed any real number θ for which there exists an infinite sequence of distinct rationals p_j/q_j satisfying $|\theta - p_j/q_j| < 1/q_j^{\omega_j}$, where $\omega_j \to \infty$ as $j \to \infty$, will be transcendental. For instance, this will hold for any infinite decimal in which there occur sufficiently long blocks of zeros or any continued fraction in which the partial quotients increase sufficiently rapidly.

There have been some remarkable improvements on Liouville's theorem, beginning with a famous work of Thue in 1909. He showed that for any algebraic number α with degree $n > 1$ and for any $\kappa > \frac{1}{2}n + 1$ there exists $c = c(\alpha, \kappa) > 0$ such that $|\alpha - p/q| > c/q^{\kappa}$ for all rationals p/q $(q > 0)$. The condition on κ was relaxed by Siegel in 1921 to $\kappa > 2\sqrt{n}$ and it was further relaxed by Dyson and Gelfond, independently, in 1947 to $\kappa > \sqrt{(2n)}$. Finally Roth proved in 1955 that it is enough to take $\kappa > 2$, and this is plainly best possible. There is an intimate connection between such results and the theory of Diophantine equations (see Chapter 8). In this context it is important to know whether the numbers $c(\alpha, \kappa)$ can be evaluated explicitly, that is, whether the results are effective. In fact all the improvements on Liouville's theorem referred to above are, in that sense, ineffective; for they involve a hypothetical assumption, made at the outset, that the inequalities in question have at least one large solution. Nevertheless effective results have been successfully obtained for particular algebraic numbers; for instance Baker proved in 1964 from properties of hypergeometric functions that, for all rationals p/q ($q>0$), we have

$$|\sqrt[3]{2}-p/q|>10^{-6}/q^{2.955}.$$

Moreover, a small but general effective improvement on Liouville's theorem, that is, valid for any algebraic α, has been established by way of the theory of linear forms in logarithms referred to in the next section.

6 Transcendental numbers

In 1873 Hermite began a new era in number theory when he succeeded in proving that e, the natural base for logarithms, is transcendental. It had earlier been established that e was neither rational nor quadratic irrational; indeed the continued fraction for e was known, namely

$$e=[2, 1, 2, 1, 1, 4, 1, 1, 6, 1, 1, 8, \ldots].$$

But Hermite's work rested on quite different ideas concerning the approximation of analytic functions by rational functions. In 1882 Lindemann found a generalization of Hermite's argument and he obtained thereby his famous proof of the transcendence of π. This sufficed to solve the ancient Greek problem of constructing, with ruler and compasses only, a square with area equal to that of a given circle. In fact, given a unit length, all the points in the plane that are capable of construction are given by the intersection of lines and circles, whence their co-ordinates in a suitable frame of reference are algebraic numbers. Hence the transcendence of π implies that the length $\sqrt{\pi}$ cannot be classically constructed and so the quadrature of the circle is impossible. Lindemann actually proved that for any distinct algebraic numbers $\alpha_1, \ldots, \alpha_n$ and any non-zero algebraic numbers $\beta_1, \ldots, \beta_n$ we have

$$\beta_1 e^{\alpha_1}+\cdots+\beta_n e^{\alpha_n} \neq 0.$$

The transcendence of π follows in view of Euler's identity $e^{i\pi}=-1$; and the result plainly includes also the transcendence of e, of $\log \alpha$ for algebraic α not 0 or 1, and of the trigonometrical functions $\cos \alpha$, $\sin \alpha$ and $\tan \alpha$ for all non-zero algebraic α.

In the sense of Lebesgue measure, 'almost all' numbers are transcendental; in fact as Cantor observed in 1874, the set of all algebraic numbers is countable. However, it has proved notoriously difficult to demonstrate the transcendence of particular numbers; for instance, Euler's constant γ has resisted any attack, and the same applies to the values $\zeta(2n+1)$ $(n=1, 2, \ldots)$ of the Riemann zeta-function, though Apéry has recently demonstrated that $\zeta(3)$ is irrational. In 1900, Hilbert raised as the seventh of his famous list of 23 problems, the question of proving the transcendence of $2^{\sqrt{2}}$ and, more generally, that of α^β for algebraic α not 0 or 1 and algebraic irrational β. Hilbert expressed the opinion that a solution lay farther in the future than the Riemann hypothesis or Fermat's last theorem. But remarkably, in 1929, following studies on integral integer-valued functions, Gelfond succeeded in verifying the special case that $e^\pi = (-1)^{-i}$ is transcendental, and a complete solution to Hilbert's seventh problem was established by Gelfond and Schneider independently in 1934. A generalization of the Gelfond–Schneider theorem was obtained by Baker in 1966; this furnished, for instance, the transcendence of $e^{\beta_0}\alpha_1^{\beta_1} \cdots \alpha_n^{\beta_n}$, and indeed that of any non-vanishing linear form

$$\beta_1 \log \alpha_1 + \cdots + \beta_n \log \alpha_n,$$

where the αs and βs denote non-zero algebraic numbers. The work enabled quantitative versions of the results to be established, giving positive lower bounds for linear forms in logarithms, and these have played a crucial role in the effective solution of a wide variety of Diophantine problems. We have already referred to one such application at the end of § 5; we shall mention some others later.

Several classical functions, apart from e^z, have been shown to assume transcendental values at non-zero algebraic values of the argument; these include the Weierstrass elliptic function $\wp(z)$, the Bessel function $J_0(z)$ and the elliptic modular function $j(z)$, where, in the latter case, z is necessarily neither real nor imaginary quadratic. In fact there is now a rich and fertile theory relating to the transcendence and algebraic independence of values assumed by analytic functions, and we refer to § 8 for an introduction to the literature.

To illustrate a few of the basic techniques of the theory, we give now a short proof of the transcendence of e; the argument can be extended quite easily to furnish the transcendence of π and indeed the general Lindemann theorem. The proof depends on properties of the integral

$$I(t)=\int_0^t e^{t-x}f(x)\,dx,$$

defined for $t\geq 0$, where f is a real polynomial with degree m. By integration by parts we have

$$I(t)=e^t\sum_{j=0}^{m} f^{(j)}(0)-\sum_{j=0}^{m} f^{(j)}(t),$$

where $f^{(j)}(x)$ denotes the jth derivative of $f(x)$. Further we observe that, if $\bar{f}$ denotes the polynomial obtained from f by replacing each coefficient with its absolute value, then

$$|I(t)|\leq\int_0^t |e^{t-x}f(x)|\,dx\leq t\,e^t\bar{f}(t).$$

Suppose now that e is algebraic, so that

$$a_0+a_1e+\cdots+a_n\,e^n=0$$

for some integers $a_0, a_1, \ldots, a_n$ with $a_0\neq 0$. We put

$$f(x)=x^{p-1}(x-1)^p\cdots(x-n)^p,$$

where p is a large prime; then the degree m of f is $(n+1)p-1$. We shall compare estimates for

$$J=a_0I(0)+a_1I(1)+\cdots+a_nI(n).$$

By the above equations we see that

$$J=-\sum_{j=0}^{m}\sum_{k=0}^{n} a_k f^{(j)}(k).$$

Now, when $1\leq k\leq n$, we have $f^{(j)}(k)=0$ for $j<p$, and

$$f^{(j)}(k)=\binom{j}{p}p!g^{(j-p)}(k)$$

for $j\geq p$, where $g(x)=f(x)/(x-k)^p$. Thus, for all j, $f^{(j)}(k)$ is an integer divisible by $p!$. Further, we have $f^{(j)}(0)=0$ for $j<p-1$, and

$$f^{(j)}(0)=\binom{j}{p-1}(p-1)!h^{(j-p+1)}(0)$$

for $j\geq p-1$, where $h(x)=f(x)/x^{p-1}$. Clearly $h^{(j)}(0)$ is an integer divisible by p for $j>0$, and $h(0)=(-1)^{np}(n!)^p$. Thus, for $j\neq p-1$, $f^{(j)}(0)$ is an integer divisible by $p!$, and $f^{(p-1)}(0)$ is an integer divisible by $(p-1)!$ but not by p for $p>n$. It follows that J is a non-zero integer divisible by $(p-1)!$, whence $|J|\geq(p-1)!$. On the other hand, the trivial estimates $\bar{f}(k)\leq(2n)^m$ and $m\leq 2np$ give

$$|J|\leq|a_1|\mathrm{e}\bar{f}(1)+\cdots+|a_n|n\,\mathrm{e}^n\bar{f}(n)\leq c^p$$

for some c independent of p. The inequalities are inconsistent for p sufficiently large, and the contradiction shows that e is transcendental, as required.

7 Minkowski's theorem

Practically intuitive deductions relating to the geometry of figures in the plane, or, more generally, in Euclidean n-space, can sometimes yield results of great importance in number theory. It was Minkowski who first systematically exploited this observation and he called the resulting study the Geometry of Numbers. The most famous theorem in this context is the convex body theorem that Minkowski obtained in 1896. By a convex body we mean a bounded, open set of points in Euclidean n-space that contains $\lambda\boldsymbol{x}+(1-\lambda)\boldsymbol{y}$ for all λ with $0<\lambda<1$ whenever it contains $\boldsymbol{x}$ and $\boldsymbol{y}$. A set of points is said to be symmetric about the origin if it contains $-\boldsymbol{x}$ whenever it contains $\boldsymbol{x}$. The simplest form of Minkowski's theorem asserts that if a convex body $\mathscr{S}$, symmetric about the origin, has volume exceeding 2^n then it contains an integer point other than the origin. By an integer point we mean a point all of whose co-ordinates are integers.

For the proof, it will suffice to verify the following result due to Blichfeldt: any bounded region $\mathscr{R}$ with volume V exceeding 1 contains distinct points $\boldsymbol{x}$, $\boldsymbol{y}$ such that $\boldsymbol{x}-\boldsymbol{y}$ is an integer point. Minkowski's theorem follows on taking $\mathscr{R}=\frac{1}{2}\mathscr{S}$, that is the set of points $\frac{1}{2}\boldsymbol{x}$ with $\boldsymbol{x}$ in $\mathscr{S}$, and noting that if $\boldsymbol{x}$ and $\boldsymbol{y}$ belong to $\mathscr{R}$ then $2\boldsymbol{x}$ and $-2\boldsymbol{y}$ belong to $\mathscr{S}$, whence $\boldsymbol{x}-\boldsymbol{y}=\frac{1}{2}(2\boldsymbol{x}-2\boldsymbol{y})$ also belongs to $\mathscr{S}$. To prove Blichfeldt's result, we note that $\mathscr{R}$ is the union of disjoint subsets $\mathscr{R}\boldsymbol{u}$, where $\boldsymbol{u}=(u_1,\ldots,u_n)$ runs through all integer points and $\mathscr{R}\boldsymbol{u}$ denotes the part of $\mathscr{R}$ that lies in the interval $u_j\leq x_j<u_j+1$ $(1\leq j\leq n)$. Thus $V=\sum V\boldsymbol{u}$, where $V\boldsymbol{u}$ denotes the volume of $\mathscr{R}\boldsymbol{u}$, and, by hypothesis, we obtain $\sum V\boldsymbol{u}>$

1. It follows that if each of the regions $\mathscr{R}\boldsymbol{u}$ is translated by $-\boldsymbol{u}$ so as to lie in the interval $0 \leq x_j < 1$ $(1 \leq j \leq n)$, then at least two of the translated regions, say the translates of $\mathscr{R}\boldsymbol{u}$ and $\mathscr{R}\boldsymbol{v}$, must overlap. Hence there exist points $\boldsymbol{x}$ in $\mathscr{R}\boldsymbol{u}$ and $\boldsymbol{y}$ in $\mathscr{R}\boldsymbol{v}$ such that $\boldsymbol{x} - \boldsymbol{u} = \boldsymbol{y} - \boldsymbol{v}$, and so $\boldsymbol{x} - \boldsymbol{y}$ is an integer point, as required.

In order to state the more general form of Minkowski's theorem we need the concept of a lattice. First we recall that points $\boldsymbol{a}_1, \ldots, \boldsymbol{a}_n$ in Euclidean n-space are said to be linearly independent if the only real numbers $t_1, \ldots, t_n$ satisfying $t_1\boldsymbol{a}_1 + \cdots + t_n\boldsymbol{a}_n = 0$ are $t_1 = \cdots = t_n = 0$; this is equivalent to the condition that $d = \det(a_{ij}) \neq 0$, where $\boldsymbol{a}_j = (a_{1j}, \ldots, a_{nj})$. By a lattice Λ we mean a set of points of the form

$$\boldsymbol{x} = u_1\boldsymbol{a}_1 + \cdots + u_n\boldsymbol{a}_n,$$

where $\boldsymbol{a}_1, \ldots, \boldsymbol{a}_n$ are fixed linearly independent points and $u_1, \ldots, u_n$ run through all the integers. The determinant of Λ is defined as $d(\Lambda) = |d|$. With this notation, the general Minkowski theorem asserts that if, for any lattice Λ, a convex body $\mathscr{S}$, symmetric about the origin, has volume exceeding $2^n d(\Lambda)$, then it contains a point of Λ other than the origin. The result can be established by simple modifications to the earlier arguments.

As an immediate application, let $\lambda_1, \ldots, \lambda_n$ be positive numbers and let $\mathscr{S}$ be the convex body $|x_j| < \lambda_j$ $(1 \leq j \leq n)$; the volume of $\mathscr{S}$ is $2^n\lambda_1 \cdots \lambda_n$. Thus, on writing

$$L_j = u_1 a_{j1} + \cdots + u_n a_{jn} \qquad (1 \leq j \leq n),$$

we deduce that if $\lambda_1 \cdots \lambda_n > d(\Lambda)$ then there exist integers $u_1, \ldots, u_n$, not all 0, such that $|L_j| < \lambda_j$ $(1 \leq j \leq n)$. This is referred to as Minkowski's linear forms theorem. It can be sharpened slightly to show that if $\lambda_1 \cdots \lambda_n = d(\Lambda)$ then there exist integers $u_1, \ldots, u_n$, not all 0, such that $|L_1| \leq \lambda_1$ and $|L_j| < \lambda_j$ $(2 \leq j \leq n)$. In fact, for each $m = 1, 2, \ldots$ there exists a non-zero integer point $\boldsymbol{u}_m$ for which $|L_1| < \lambda_1 + 1/m$ and $|L_j| < \lambda_j$ $(2 \leq j \leq n)$; but the $\boldsymbol{u}_m$ are bounded, and so $\boldsymbol{u}_m = \boldsymbol{u}$ for some fixed $\boldsymbol{u}$ and infinitely many m, whence $\boldsymbol{u} = (u_1, \ldots, u_n)$ has the required properties.

Minkowski's linear forms theorem implies that if $\theta_1, \ldots, \theta_n$ are any real numbers and if $Q > 0$ then there exist integers p, $q_1, \ldots, q_n$, not all 0, such that $|q_j| < Q$ $(1 \leq j \leq n)$ and

$$|q_1\theta_1 + \cdots + q_n\theta_n - p| \leq Q^{-n}.$$

Similarly we see that there exist integers $p_1, \ldots, p_n, q$, not all 0, such that $|q| \le Q^n$ and $|q\theta_j - p_j| < 1/Q$ $(1 \le j \le n)$. It follows that, if one at least of $\theta_1, \ldots, \theta_n$ is irrational, then

$$|\theta_j - p_j/q| < q^{-1-1/n} \qquad (1 \le j \le n)$$

for infinitely many rationals p_j/q $(q > 0)$. These results generalize Dirichlet's theorem discussed in § 1. In the opposite direction, it is easy to extend the observation on quadratic irrationals made in § 5 to show that, when θ is an algebraic number with degree $n+1$, there exists $c = c(\theta) > 0$ such that

$$|q_1\theta + \cdots + q_n\theta^n - p| > cq^{-n}$$

for all integers $p, q_1, \ldots, q_n$ with $q = \max|q_j| > 0$. This implies, by a classical transference principle, that the exponent $-1-1/n$ above is best possible. It is known from transcendence theory that, for any $\varepsilon > 0$, there exists $c > 0$ such that

$$|q_1 \mathrm{e} + \cdots + q_n \mathrm{e}^n - p| > cq^{-n-\varepsilon}$$

for all integers $p, q_1, \ldots, q_n$ with $q = \max|q_j| > 0$. Moreover, some deep work of Schmidt, generalizing the Thue–Siegel–Roth theorem, shows that the same holds when $\mathrm{e}, \ldots, \mathrm{e}^n$ are replaced by algebraic numbers $\theta_1, \ldots, \theta_n$ with $1, \theta_1, \ldots, \theta_n$ linearly independent over the rationals; in analogy with lattice points, we say that real numbers $\phi_1, \ldots, \phi_m$ are linearly independent over the rationals if the only rationals $t_1, \ldots, t_m$ satisfying $t_1\phi_1 + \cdots + t_m\phi_m = 0$ are $t_1 = \cdots = t_m = 0$.

Minkowski conjectured that if $L_1, \ldots, L_n$ are linear forms as above and if $\theta_1, \ldots, \theta_n$ are any real numbers then there exist integers $u_1, \ldots, u_n$ such that

$$|L_1 - \theta_1| \cdots |L_n - \theta_n| \le 2^{-n} d(\Lambda).$$

At present the conjecture remains open, but it is trivial in the case $n = 1$ and Minkowski himself proved that it is valid in the case $n = 2$. It has subsequently been verified for $n = 3, 4$ and 5, and Tchebotarev showed that it holds for all n if 2^{-n} is replaced by $2^{-\frac{1}{2}n}$. Minkowski's work furnished a result to the effect that if θ is irrational and θ' is not of the form $r\theta + s$ for integers r, s, then there are infinitely many integers $q \ne 0$ such that, for some integer p,

$$|q\theta - p - \theta'| < 1/(4|q|);$$

and here the constant 1/4 is best possible. The result implies that the numbers $\{n\theta\}$, where $n=1, 2, \ldots$, are dense in the unit interval, that is, for every θ' with $0<\theta'<1$, and for every $\varepsilon>0$, we have $|\{n\theta\}-\theta'|<\varepsilon$ for some n. A famous theorem of Kronecker implies that, more generally, the points

$$(\{n\theta_1\}, \ldots, \{n\theta_m\}) \qquad (n=1, 2, \ldots),$$

where $1, \theta_1, \ldots, \theta_m$ are linearly independent over the rationals, are dense in the unit cube in Euclidean m-space.

8 Further reading

The classic text on continued fractions is Perron's *Die Lehre von den Kettenbrüchen* (Teubner, Leipzig, 1913). There are, however, useful accounts in most introductory works on number theory; see, in particular, Cassels' *An introduction to Diophantine approximation* (Cambridge U.P., 1957), and the books of Niven and Zuckerman (Wiley, New York, 1966) and of Hardy and Wright (Oxford, 5th edn, 1979) cited earlier. A nice, short work is Khintchine's *Kettenbrüche* (Teubner, Leipzig, 1956).

Numerous references to the literature relating to § 5 and § 6 are given in Baker's *Transcendental number theory* (Cambridge U.P., 2nd edn, 1979). For advanced work concerning rational approximations to algebraic numbers see W. M. Schmidt, *Diophantine approximation* (Springer Math. Lecture Notes, 785, Berlin 1980). The topics referred to in § 7 are discussed fully in Cassels' *An introduction to the geometry of numbers* (Springer-Verlag, Berlin, 1971).

9 Exercises

(i) Evaluate the continued fraction $[1, 2, 3, \overline{1, 4}]$.

(ii) Assuming that π is given by $3.141\,592\,6\ldots$, correct to seven decimal places, prove that the first three convergents to π are $\frac{22}{7}$, $\frac{333}{106}$ and $\frac{355}{113}$. Verify that $|\pi-\frac{355}{113}|<10^{-6}$.

(iii) Let θ, θ' be the roots of the equation $x^2-ax-1=0$, where a is a positive integer and $\theta>0$. Show that the denominators in the convergents to θ are given by

$q_{n-1} = (\theta^n - \theta'^n)/(\theta - \theta')$. Verify that the Fibonacci sequence 1, 1, 2, 3, 5, ... is given by $q_0, q_1, \ldots$ in a special case.

(iv) Prove that the denominators q_n in the convergents to any real θ satisfy $q_n \geq (\frac{1}{2}(1+\sqrt{5}))^{n-1}$. Prove also that, if the partial quotients are bounded above by a constant A, then $q_n \leq (\frac{1}{2}(A+\sqrt{(A^2+4)}))^n$.

(v) Assuming that the continued fraction for e is as quoted in § 6, show that $|e - p/q| > c/(q^2 \log q)$ for all rationals p/q $(q > 1)$, where c is a positive constant.

(vi) Assuming the Thue–Siegel–Roth theorem, show that the sum $a^{-b} + a^{-b^2} + a^{-b^3} + \ldots$ is transcendental for any integers $a \geq 2$, $b \geq 3$.

(vii) Let $\alpha, \beta, \gamma, \delta$ be real numbers with $\Delta = \alpha\delta - \beta\gamma \neq 0$. Prove that there exist integers x, y, not both 0, such that $|L| + |M| \leq (2|\Delta|)^{1/2}$, where $L = \alpha x + \beta y$ and $M = \gamma x + \delta y$. Deduce that the inequality $|LM| \leq \frac{1}{2}|\Delta|$ is soluble non-trivially.

(viii) With the same notation, prove that the inequality $L^2 + M^2 \leq (4/\pi)|\Delta|$ is soluble in integers x, y, not both 0. Verify that the constant $4/\pi$ cannot be replaced by a number smaller than $(4/3)^{\frac{1}{2}}$.

(ix) Assuming Kronecker's theorem and the transcendence of e^π, show that, for any primes $p_1, \ldots, p_m$, there exists an integer $n > 0$ such that $\cos(\log p_j^n) \leq -\frac{1}{2}$ for $j = 1, 2, \ldots, m$.

7

Quadratic fields

1 Algebraic number fields

Although we shall be concerned in the sequel only with quadratic fields, we shall nevertheless begin with a short discussion of the more general concept of an algebraic number field. The theory relating to such fields has arisen from attempts to solve Fermat's last theorem and it is one of the most beautiful and profound in mathematics.

Let α be an algebraic number with degree n and let P be the minimal polynomial for α (see § 5 of Chapter 6). By the conjugates of α we mean the zeros $\alpha_1, \ldots, \alpha_n$ of P. The algebraic number field k generated by α over the rationals $\mathbb{Q}$ is defined as the set of numbers $Q(\alpha)$, where $Q(x)$ is any polynomial with rational coefficients; the set can be regarded as being embedded in the complex number field $\mathbb{C}$ and thus its elements are subject to the usual operations of addition and multiplication. To verify that k is indeed a field we have to show that every non-zero element $Q(\alpha)$ has an inverse. Now, if P is the minimal polynomial for α as above, then P, Q are relatively prime and so there exist polynomials R, S such that $PS+QR=1$ identically, that is for all x. On putting $x=\alpha$ this gives $R(\alpha)=1/Q(\alpha)$, as required. The field k is said to have degree n over $\mathbb{Q}$, and one writes $[k:\mathbb{Q}]=n$.

The construction can be continued analogously to furnish, for every algebraic number field k and every algebraic number β, a field $K=k(\beta)$ with elements given by polynomials in β with coefficients in k. The degree $[K:k]$ of K over k is defined in the obvious way as the degree of β over k. Now K is in fact an algebraic number field over $\mathbb{Q}$, for it can be shown that $K=\mathbb{Q}(\gamma)$, where $\gamma = u\alpha + v\beta$ for some rationals u, v; thus we have

$$[K:k][k:\mathbf{Q}] = [K:\mathbf{Q}].$$

An algebraic number is said to be an algebraic integer if the coefficient of the highest power of x in the minimal polynomial P is 1. The algebraic integers in an algebraic number field k form a ring R. The ring has an integral basis, that is, there exist elements $\omega_1, \ldots, \omega_n$ in R such that every element in R can be expressed uniquely in the form $u_1\omega_1 + \cdots + u_n\omega_n$ for some rational integers $u_1, \ldots, u_n$. We can write $\omega_i = p_i(\alpha)$, where p_i denotes a polynomial over $\mathbf{Q}$, and it is then readily verified that the number $(\det p_i(\alpha_j))^2$ is a rational integer independent of the choice of basis; it is called the discriminant of k and it turns out to be an important invariant.

An algebraic integer α is said to be divisible by an algebraic integer β if α/β is an algebraic integer. An algebraic integer ε is said to be a unit if $1/\varepsilon$ is an algebraic integer. Suppose now that R is the ring of algebraic integers in a number field k. Two elements α, β of R are said to be associates if $\alpha = \varepsilon\beta$ for some unit ε, and this is an equivalence relation on R. An element α of R is said to be irreducible if every divisor of α in R is either an associate or a unit. One calls R a unique factorization domain if every element of R can be expressed essentially uniquely as a product of irreducible elements. The fundamental theorem of arithmetic asserts that the ring of integers in $k = \mathbf{Q}$ has this property; but it does not hold for every k. Nevertheless, it is known from pioneering studies of Kummer and Dedekind that a unique factorization property can be restored by the introduction of ideals, and this forms the central theme of algebraic number theory. The work on Fermat's last theorem that motivated much of the subject related to the particular case of the cyclotomic field $\mathbf{Q}(\zeta)$ where ζ is a root of unity.

2 The quadratic field

Let d be a square-free integer, positive or negative, but not 1. The quadratic field $\mathbf{Q}(\sqrt{d})$ is the set of all numbers of the form $u + v\sqrt{d}$ with rational u, v, subject to the usual operations of addition and multiplication. For any element $\alpha = u + v\sqrt{d}$ in $\mathbf{Q}(\sqrt{d})$ one defines the norm of α as the rational number $N(\alpha) =$

u^2-dv^2. Clearly $N(\alpha)=\alpha\bar{\alpha}$, where $\bar{\alpha}=u-v\sqrt{d}$; $\bar{\alpha}$ is called the conjugate of α. Now for any elements α, β in $\mathbb{Q}(\sqrt{d})$ we see that $\overline{\alpha\beta}=\bar{\alpha}\bar{\beta}$ and thus we have the important formula $N(\alpha)N(\beta)=N(\alpha\beta)$. It is readily verified that $\mathbb{Q}(\sqrt{d})$ is indeed a field; in particular, the inverse of any non-zero element α is $\bar{\alpha}/N(\alpha)$. The special field $\mathbb{Q}(\sqrt{(-1)})$ is called the Gaussian field and it is customary to express its elements in the form $u+iv$; in this case we have $N(\alpha)=u^2+v^2$ and so the product formula is precisely the identity referred to in § 4 of Chapter 5.

We proceed now to determine the algebraic integers in $\mathbb{Q}(\sqrt{d})$. Suppose that $\alpha=u+v\sqrt{d}$ is such an integer and let $a=2u$, $b=2v$. Then α is a zero of the polynomial $P(x)=x^2-ax+c$, where $c=N(\alpha)$, and so the rational numbers a, c must in fact be integers. We have $4c=a^2-b^2d$ and, since d is square-free, it follows that also b is a rational integer. Now if $d\equiv 2$ or $3 \pmod 4$ then, since a square is congruent to 0 or $1 \pmod 4$, we see that a, b are even and thus u, v are rational integers; hence, in this case, an integral basis for $\mathbb{Q}(\sqrt{d})$ is given by $1, \sqrt{d}$. If $d\equiv 1 \pmod 4$, which is the only other possibility, then $a\equiv b \pmod 2$ and thus $u-v$ is a rational integer; recalling that $v=\frac{1}{2}b$, we conclude that, in this case, an integral basis for $\mathbb{Q}(\sqrt{d})$ is given by $1, \frac{1}{2}(1+\sqrt{d})$. The discriminant D of $\mathbb{Q}(\sqrt{d})$, as defined in § 1, is therefore $4d$ when $d\equiv 2$ or $3 \pmod 4$ and it is d when $d\equiv 1 \pmod 4$.

There is a close analogy between the theory of quadratic fields and the theory of binary quadratic forms as described in Chapter 5. In particular, the discriminant D of $\mathbb{Q}(\sqrt{d})$ is congruent to 0 or $1 \pmod 4$ and so D is also the discriminant of a binary quadratic form. Now if α is any algebraic integer in $\mathbb{Q}(\sqrt{d})$ then, for some rational integers x, y, we have $\alpha=x+y\sqrt{d}$ when $d\equiv 2$ or $3 \pmod 4$ and $\alpha=x+\frac{1}{2}y(1+\sqrt{d})$ when $d\equiv 1 \pmod 4$. Thus we see that $N(\alpha)=F(x,y)$, where F denotes the principal form with discriminant D, that is x^2-dy^2 when $D\equiv 0 \pmod 4$ and $(x+\frac{1}{2}y)^2-\frac{1}{4}dy^2$ when $D\equiv 1 \pmod 4$.

3 Units

By a unit in $\mathbb{Q}(\sqrt{d})$ we mean an algebraic integer ε in $\mathbb{Q}(\sqrt{d})$ such that $1/\varepsilon$ is an algebraic integer. Plainly if ε is a unit then $N(\varepsilon)$ and $N(1/\varepsilon)$ are rational integers and, since $N(\varepsilon)N(1/\varepsilon)=1$, we see that $N(\varepsilon)=\pm 1$. Conversely, if $N(\varepsilon)=\pm 1$, then $\varepsilon\bar{\varepsilon}=\pm 1$ and so ε is a unit. Thus, by the above remarks, the units in $\mathbf{Q}(\sqrt{d})$ are determined by the integer solutions x, y of the equation $F(x, y)=\pm 1$.

We shall distinguish two cases according as $d<0$ or $d>0$; in the first case the quadratic field is said to be imaginary and in the second it is said to be real. Now in an imaginary quadratic field there are only finitely many units. In fact if $D<-4$ then, as is readily verified, the equation $F(x, y)=\pm 1$ has only the solutions $x=\pm 1$, $y=0$ and so the only units in $\mathbf{Q}(\sqrt{d})$ are ± 1. For $d=-1$, that is, for the Gaussian field, we have $F(x, y)=x^2+y^2$ and there are therefore four units ± 1, $\pm \mathrm{i}$. For $d=-3$ we have $F(x, y)=x^2+xy+y^2$ and the equation $F(x, y)=\pm 1$ has six solutions, namely $(\pm 1, 0)$, $(0, \pm 1)$, $(1, -1)$ and $(-1, 1)$; thus the units of $\mathbf{Q}(\sqrt{(-3)})$ are ± 1, $\frac{1}{2}(\pm 1 \pm\sqrt{(-3)})$. It follows that the units in an imaginary quadratic field are all roots of unity; they are given by the zeros of x^2-1 when $D<-4$, by those of x^4-1 when $D=-4$ and by those of x^6-1 when $D=-3$. Hence the number of units is the same as the number w for forms with discriminant D indicated in § 3 of Chapter 5.

We turn now to real quadratic fields; in this case there are infinitely many units. To establish the result it suffices to show that, when $d>0$, there is a unit η in $\mathbf{Q}(\sqrt{d})$ other than ± 1; for then η^m is a unit for all integers m, and, since the only roots of unity in $\mathbf{Q}(\sqrt{d})$ are ± 1, we see that different m give distinct units. We shall use Dirichlet's theorem on Diophantine approximation (see § 1 of Chapter 6); the theorem implies that, for any integer $Q>1$, there exist rational integers p, q, with $0<q<Q$, such that $|\alpha|\leq 1/Q$, where $\alpha=p-q\sqrt{d}$. Now the conjugate $\bar{\alpha}=\alpha+2q\sqrt{d}$ satisfies $|\bar{\alpha}|\leq 3Q\sqrt{d}$ and thus we have $|N(\alpha)|\leq 3\sqrt{d}$. Further, since $\sqrt{d}$ is irrational, we obtain, as $Q\to\infty$, infinitely many α with this property. But $N(\alpha)$ is a rational integer bounded independently of Q, and thus, for infinitely many α, it takes some fixed value, say N. Moreover we can select two distinct elements from the infinite set, say $\alpha=p-q\sqrt{d}$ and $\alpha'=p'-q'\sqrt{d}$, such that $p\equiv p' \pmod{N}$ and $q\equiv q' \pmod{N}$. We now put $\eta=\alpha/\alpha'$. Then $N(\eta)=N(\alpha)/N(\alpha')=1$. Further, η is clearly not 1,

and it is also not -1 since $\sqrt{d}$ is irrational and q, q' are positive. Furthermore we have $\eta = x + y\sqrt{d}$, where $x = (pp' - dqq')/N$ and $y = (pq' - p'q)/N$, and the congruences above imply that x, y are rational integers. Thus η is a non-trivial unit in $\mathbb{Q}(\sqrt{d})$, as required. The argument here shows, incidentally, that the Pell equation $x^2 - dy^2 = 1$ has a non-trivial solution; we shall discuss the equation more fully in Chapter 8.

We can now give a simple expression for all the units in a real quadratic field. In fact consider the set of all units in the field that exceed 1. The set is not empty, for if η is the unit obtained above then one of the numbers $\pm\eta$ or $\pm 1/\eta$ is a member. Further, each element of the set has the form $u + v\sqrt{d}$, where u, v are either integers, or, in the case $d \equiv 1 \pmod 4$, possibly halves of odd integers. Furthermore u and v are positive, for $u + v\sqrt{d}$ is greater than its conjugate $u - v\sqrt{d}$, which lies between -1 and 1. It follows that there is a smallest element in the set, say ε. Now if ε' is any positive unit in the field then there is a unique integer m such that $\varepsilon^m \leq \varepsilon' < \varepsilon^{m+1}$; this gives $1 \leq \varepsilon'/\varepsilon^m < \varepsilon$. But $\varepsilon'/\varepsilon^m$ is also a unit in the field and thus, from the definition of ε, we conclude that $\varepsilon' = \varepsilon^m$. This shows that all the units in the field are given by $\pm\varepsilon^m$, where $m = 0, \pm 1, \pm 2, \ldots$.

The results established here for quadratic fields are special cases of a famous theorem of Dirichlet concerning units in an arbitrary algebraic number field. Suppose that the field k is generated by an algebraic number α with degree n and that precisely s of the conjugates $\alpha_1, \ldots, \alpha_n$ of α are real; then $n = s + 2t$, where t is the number of complex conjugate pairs. Dirichlet's theorem asserts that there exist $r = s + t - 1$ fundamental units $\varepsilon_1, \ldots, \varepsilon_r$ in k such that every unit in k can be expressed uniquely in the form $\rho\varepsilon_1^{m_1} \cdots \varepsilon_r^{m_r}$, where $m_1, \ldots, m_r$ are rational integers and ρ is a root of unity in k.

4 Primes and factorization

Let R be the ring of algebraic integers in a quadratic field $\mathbb{Q}(\sqrt{d})$. By a prime π in R we mean an element of R that is neither 0 nor a unit and which has the property that if π divides $\alpha\beta$, where α, β are elements of R, then either π divides α or π divides β. It will be noted at once that a prime π is irreducible in the sense indicated in § 1; for if $\pi = \alpha\beta$ then either α/π or β/π is an element of R whence either β or α is a unit. However, an irreducible element need not be a prime. Consider, for example, the number 2 in the quadratic field $\mathbb{Q}(\sqrt{(-5)})$. It is certainly irreducible, for if $2 = \alpha\beta$ then $4 = N(\alpha)N(\beta)$; but $N(\alpha)$ and $N(\beta)$ have the form x^2+5y^2 for some integers x, y, and, since the equation $x^2+5y^2 = \pm 2$ has no integer solutions, it follows that either $N(\alpha) = \pm 1$ or $N(\beta) = \pm 1$ and thus either α or β is a unit. On the other hand, 2 is not a prime in $\mathbb{Q}(\sqrt{(-5)})$, for it divides

$$(1+\sqrt{(-5)})(1-\sqrt{(-5)}) = 6,$$

but it does not divide either $1+\sqrt{(-5)}$ or $1-\sqrt{(-5)}$; indeed, by taking norms, it is readily verified that each of the latter is irreducible.

Now every element α of R that is neither 0 nor a unit can be factorized into a finite product of irreducible elements. For if α is not itself irreducible than $\alpha = \beta\gamma$ for some β, γ in R, neither of which is a unit. If β were not irreducible then it could be factorized likewise, and the same holds for γ. The process must terminate, for if $\alpha = \beta_1 \cdots \beta_n$, where none of the βs is a unit, then, since $|N(\beta_j)| \geq 2$, we see that $|N(\alpha)| \geq 2^n$. The ring R is said to be a unique factorization domain if the expression for α as a finite product of irreducible elements is essentially unique, that is, unique except for the order of the factors and the possible replacement of irreducible elements by their associates. A fundamental problem in number theory is to determine which domains have unique factorization, and here the definition of a prime plays a crucial role. In fact we have the basic theorem that R is a unique factorization domain if and only if every irreducible element of R is also a prime in R. To verify the assertion, note that if factorization in R is unique and if π is an irreducible element such that π divides $\alpha\beta$ for some α, β in R then π must be an associate of one of the irreducible factors of α or β and so π divides α or β, as required. Conversely, if every irreducible element is also a prime then we can argue as in the demonstration of the fundamental theorem of arithmetic given in § 5 of Chapter 1; thus if $\alpha = \pi_1 \cdots \pi_k$ as a product of irreducible elements, and if π' is an irreducible element occurring in another factorization, then π' must divide π_j for some j, whence π' and π_j are associates, and assuming by induction that the result holds for α/π', the required uniqueness of factorization follows.

All the imaginary quadratic fields $\mathbb{Q}(\sqrt{d})$ which have the unique factorization property are known; they are given by $d=-1, -2, -3, -7, -11, -19, -43, -67$ and -163. The theorem has a long history, dating back to Gauss, and it was finally proved by Baker and Stark, independently, in 1966; the methods of proof were quite different, one depending on transcendence theory (cf. § 6 of Chapter 6) and the other on the study of elliptic modular functions. The theorem shows, incidentally, that the nine discriminants d indicated in Exercise (i) of § 7 of Chapter 5 are the only values for which $h(d)=1$. The problem of finding all the real quadratic fields $\mathbb{Q}(\sqrt{d})$ with unique factorization remains open; it is generally conjectured that there are infinitely many such fields but even this has not been proved. Nevertheless all such fields with d relatively small, for instance with $d<100$, are known; we shall discuss some particular cases in the next section.

5 Euclidean fields

A quadratic field $\mathbb{Q}(\sqrt{d})$ is said to be Euclidean if its ring of integers R has the property that, for any elements α, β of R with $\beta \neq 0$, there exist elements γ, δ of R such that $\alpha=\beta\gamma+\delta$ and $|N(\delta)|<|N(\beta)|$. For such fields there exists a Euclidean algorithm analogous to that described in Chapter 1. In fact we can generate the sequence of equations $\delta_{j-2}=\delta_{j-1}\gamma_j+\delta_j$ ($j=1, 2, \ldots$), where $\delta_{-1}=\alpha$, $\delta_0=\beta$, $\delta_1=\delta$, $\gamma_1=\gamma$ and $|N(\delta_j)|<|N(\delta_{j-1})|$; the sequence terminates when $\delta_{k+1}=0$ for some k and then δ_k has the properties of a greatest common divisor, that is, δ_k divides α and β, and every common divisor of α, β divides δ_k. Moreover we have $\delta_k=\alpha\lambda+\beta\mu$ for some λ, μ in R. This can be verified either by successive substitution or by observing that $|N(\delta_k)|$ is the least member of the set of positive integers of the form $|N(\alpha\lambda+\beta\mu)|$, where λ, μ run through the elements of R. In fact the set certainly has a least member $|N(\delta')|$, say, where $\delta'=\alpha\lambda+\beta\mu$ for some λ, μ in R; thus every common divisor of

α, β divides δ'. Further, δ' divides α, since from $\alpha=\delta'\gamma+\delta''$, with $|N(\delta'')|<|N(\delta')|$, we see that $\delta''=\alpha\lambda'+\beta\mu'$ for some λ', μ' in R, whence $N(\delta'')=0$ and so $\delta''=0$; similarly δ' divides β. Hence we have $\delta'=\delta_k$. It is clear that if δ_k is a unit then, by division, we obtain elements λ, μ in R with $\alpha\lambda+\beta\mu=1$.

We proceed now to prove that a Euclidean field has unique factorization. It suffices, in view of § 4, to show that every irreducible element π in R is a prime; accordingly suppose that π divides $\alpha\beta$ but that π does not divide α. Then, by the remarks above, there exist integers λ, μ in R such that $\alpha\lambda+\pi\mu=1$. This gives $\alpha\beta\lambda+\pi\beta\mu=\beta$, whence π divides β. Thus π is a prime and the desired result follows.

It was proved by Chatland and Davenport in 1950, and independently by Inkeri at about the same time, that there are precisely 21 Euclidean fields $\mathbb{Q}(\sqrt{d})$; the values of d are given by $-11, -7, -3, -2, -1, 2, 3, 5, 6, 7, 11, 13, 17, 19, 21, 29, 33, 37, 41, 57$ and 73. It had been proved earlier by Heilbronn that the list must be finite, and it had been verified as a consequence of works by Dickson, Perron, Oppenheimer, Remak and Rédei that the fields listed here are indeed Euclidean; we shall confirm the assertion for the first eight fields in a moment. It is easy to see that there can be no other Euclidean fields with $d<0$. In fact if $d\equiv 2$ or $3 \pmod 4$ and $d\le -5$ then we cannot have $\sqrt{d}=2\gamma+\delta$ with $|N(\delta)|<4$; for we can express γ, δ as $x+y\sqrt{d}$ and $x'+y'\sqrt{d}$ respectively, where x, y and x', y' are rational integers, and since $N(\delta)\ge x'^2+5y'^2$ we would obtain $y'=0$, contrary to $2y+y'=1$. Similarly if $d\equiv 1 \pmod 4$ and $d\le -15$, then we cannot have $\frac{1}{2}(1+\sqrt{d})=2\gamma+\delta$ with $|N(\delta)|<4$. The most difficult part of the theorem is the proof that there are no other Euclidean fields with $d>0$. In this connection, Davenport showed by an ingenious algorithm derived from studies on Diophantine approximation that if $d>2^{14}$ then $\mathbb{Q}(\sqrt{d})$ is not Euclidean; this reduced the problem to a finite checking of cases. Incidentally, Rédei claimed originally that the field $\mathbb{Q}(\sqrt{97})$ was Euclidean but Barnes and Swinnerton-Dyer proved this to be erroneous.

We shall show now that if $d=-2, -1, 2$ or 3 then $\mathbb{Q}(\sqrt{d})$ is Euclidean. Accordingly let α, β be any algebraic integers in $\mathbb{Q}(\sqrt{d})$ with $\beta\ne 0$. Then $\alpha/\beta=u+v\sqrt{d}$ for some rationals u, v.

We select integers x, y as close as possible to u, v and put $r = u - x$, $s = v - y$; then $|r| \leq \frac{1}{2}$ and $|s| \leq \frac{1}{2}$. On writing $\gamma = x + y\sqrt{d}$ we obtain $\alpha = \beta\gamma + \delta$, where $\delta = \beta(r + s\sqrt{d})$. This gives $N(\delta) = N(\beta)(r^2 - ds^2)$. But for $|d| \leq 2$ we have $|r^2 - ds^2| \leq r^2 + 2s^2 \leq \frac{3}{4}$, and for $d = 3$ we have $|r^2 - ds^2| \leq \max(r^2, ds^2) \leq \frac{3}{4}$. Hence $|N(\delta)| < |N(\beta)|$, as required.

Finally we prove that $\mathbb{Q}(\sqrt{d})$ is Euclidean when $d = -11, -7, -3, 5$ and 13. In these cases we have $d \equiv 1 \pmod 4$ and so 1, $\frac{1}{2}(1 + \sqrt{d})$ is an integral basis for $\mathbb{Q}(\sqrt{d})$. Again let α, β be any algebraic integers in $\mathbb{Q}(\sqrt{d})$, with $\beta \neq 0$, and let $\alpha/\beta = u + v\sqrt{d}$ with u, v rational. We select an integer y as close as possible to $2v$ and put $s = v - \frac{1}{2}y$; then $|s| \leq \frac{1}{4}$. Further we select an integer x as close as possible to $u - \frac{1}{2}y$ and put $r = u - x - \frac{1}{2}y$; then $|r| \leq \frac{1}{2}$. On writing $\gamma = x + \frac{1}{2}y(1 + \sqrt{d})$ we see that $\alpha = \beta\gamma + \delta$, where $\delta = \beta(r + s\sqrt{d})$. Now, for $|d| \leq 11$, we have $|r^2 - ds^2| \leq \frac{1}{4} + \frac{11}{16} < 1$, and, for $d = 13$, we have $|r^2 - ds^2| \leq \frac{13}{16}$. The result follows.

6 The Gaussian field

To conclude this chapter we shall describe the principal properties of the most fundamental quadratic field, namely the Gaussian field $\mathbb{Q}(\sqrt{(-1)})$ or $\mathbb{Q}(\mathrm{i})$. We have already seen that the integers in the field, that is, the Gaussian integers, have the form $x + \mathrm{i}y$ with x, y rational integers. Thus the norm of a Gaussian integer has the form $x^2 + y^2$, and, in particular, it is non-negative. It was noted in § 3 that there are just four units ± 1 and $\pm \mathrm{i}$. Moreover we proved in § 5 that the field is Euclidean and so has unique factorization. Hence there is no need to distinguish between irreducible elements and primes, and we shall use the latter terminology in preference; in fact we shall refer to the elements as Gaussian primes.

Our purpose now is to determine all the Gaussian primes. We begin with two preliminary observations which actually apply analogously to all quadratic fields with unique factorization. First, if α is any Gaussian integer and if $N(\alpha)$ is a rational prime then α is a Gaussian prime; for plainly if $\alpha = \beta\gamma$ for some Gaussian integers β, γ then $N(\alpha) = N(\beta)N(\gamma)$ and so either $N(\beta) = 1$ or $N(\gamma) = 1$ whence either β or γ is a unit. Secondly we observe that every Gaussian prime π divides just one rational prime p. For π certainly divides $N(\pi)$ and so there is a least positive rational integer p such that π divides p; and p is a rational prime, for if $p = mn$, where m, n are rational integers, then, since π is a Gaussian prime, we have either π divides m or π divides n whence, by the minimal property of p, either m or n is 1. The prime p is unique for if p' is any other rational prime then there exist rational integers a, a' such that $ap + a'p' = 1$; thus if π were to divide both p and p' then it would divide 1 and so be a unit contrary to definition.

We note next that a rational prime p is either itself a Gaussian prime or it is the product $\pi\pi'$ of two Gaussian primes, where π, π' are conjugates. Indeed p is divisible by some Gaussian prime π and thus we have $p = \pi\lambda$ for some Gaussian integer λ; this gives $N(\pi)N(\lambda) = p^2$ and the two cases correspond to the possibilities $N(\lambda) = 1$, implying that λ is a unit and that p is an associate of π, and $N(\lambda) = p$, implying that $N(\pi) = p$. Now the first case applies when $p \equiv 3 \pmod 4$ and the second when $p \equiv 1 \pmod 4$. For $N(\pi)$ has the form $x^2 + y^2$ and a square is congruent to 0 or 1 (mod 4). Further, if $p \equiv 1 \pmod 4$, then -1 is a quadratic residue (mod p) whence p divides $x^2 + 1 = (x + \mathrm{i})(x - \mathrm{i})$ for some rational integer x; but if p were a Gaussian prime then it would divide either $x + \mathrm{i}$ or $x - \mathrm{i}$, contrary to the fact that neither $x/p + \mathrm{i}/p$ nor $x/p - \mathrm{i}/p$ is a Gaussian integer. With regard to the prime 2, we have $2 = (1 + \mathrm{i})(1 - \mathrm{i})$ and here $1 + \mathrm{i}$ and $1 - \mathrm{i}$ are Gaussian primes and, moreover, associates. Combining our results, we find therefore that the totality of Gaussian primes are given by the rational primes $p \equiv 3 \pmod 4$, by the factors π, π' in the expression $p = \pi\pi'$ appertaining to primes $p \equiv 1 \pmod 4$, and by $1 + \mathrm{i}$, together with all their associates formed by multiplying with ± 1 and $\pm\mathrm{i}$. The argument here furnishes, incidentally, another proof of the result that every prime $p \equiv 1 \pmod 4$ can be expressed as a sum of two squares (see § 4 of Chapter 5).

Many of the definitions and theorems discussed earlier for the rational field possess natural analogues in the Gaussian field. Thus, for example, one can specify greatest common divisors and congruences in an obvious way, and there is an analogue of Fermat's theorem to the effect that if π is a Gaussian prime and α is a Gaussian integer, with $(\alpha, \pi) = 1$, then $\alpha^{N(\pi)-1} \equiv$

1 (mod π). There is also, for instance, an analogue of the prime number theorem to the effect that the number of non-associated Gaussian primes π with $N(\pi) \leq x$ is asymptotic to $x/\log x$ as $x \to \infty$.

7 Further reading

The structure of quadratic fields can be properly appreciated only in the wider context of algebraic number theory and with reference especially to the theory of ideals. The classic text in this connection is that of Hecke. It was originally published in German in 1923 and it has recently appeared in English translation under the title *Lectures on the theory of algebraic numbers* (Graduate texts in Mathematics, Vol. 77, Springer-Verlag, Berlin, 1981); it remains one of the best works on the subject. There are several newer expositions however. In particular the book *Algebraic number theory* by I. Stewart and D. Tall (Chapman and Hall, London, 1979) is relatively elementary and easy to read, while the volume with the same title edited by J. W. S. Cassels and A. Fröhlich (Academic Press, London, 1967) and that by S. Lang (Addison-Wesley, Reading, Mass., 1970) include accounts of more advanced topics. Some other good works are E. Artin's *Theory of algebraic numbers* (Göttingen, 1956) and W. Narkiewicz's *Elementary and analytic theory of algebraic numbers* (Polish Acad. Sci., Mon. Mat. 57, Warsaw, 1974). The book *Basic number theory* (Springer-Verlag, Berlin, 1967) by A. Weil covers similar ground but is written on a very sophisticated level.

An account of the solution, mentioned in § 4, to the problem of determining all imaginary quadratic fields with unique factorization can be found in Chapter 5 of Baker's *Transcendental number theory* (Cambridge U.P., 1979). The work, referred to in § 5, of Chatland and Davenport on Euclidean fields appeared in the *Canadian J. Math.* **2** (1950), 289–96; the article is reprinted in *The collected works of Harold Davenport* (Academic Press, London, 1978), Vol. I, pp. 366–73. For a proof of the result on Gaussian primes cited at the end of § 6 see E. Landau's *Einführung in die elementare und analytische Theorie der algebraischen Zahlen und der Ideale* (Teubner, Leipzig, 1918).

8 Exercises

(i) Show that the units in $\mathbb{Q}(\sqrt{2})$ are given by $\pm(1+\sqrt{2})^n$, where $n=0, \pm1, \pm2, \ldots$. Find the units in $\mathbb{Q}(\sqrt{3})$.

(ii) Determine the integers n and d for which $(1+n\sqrt{d})/(1-n\sqrt{d})$ is a unit in $\mathbb{Q}(\sqrt{d})$.

(iii) By considering products of norms, or otherwise, prove that there are infinitely many irreducible elements in the integral domain of any quadratic field.

(iv) Explain why the equation $2.11=(5+\sqrt{3})(5-\sqrt{3})$ is not inconsistent with the fact that $\mathbb{Q}(\sqrt{3})$ has unique factorization.

(v) Prove that the equation $2.3=(\sqrt{(-6)})(-\sqrt{(-6)})$ implies that $\mathbb{Q}(\sqrt{-6})$ does not have unique factorization.

(vi) Show that $1+\sqrt{(-17)}$ is irreducible in $\mathbb{Q}(\sqrt{(-17)})$. Verify that $\mathbb{Q}(\sqrt{(-17)})$ does not have unique factorization.

(vii) Find equations to show that $\mathbb{Q}(\sqrt{d})$ does not have unique factorization for $d=-10, -13, -14$ and -15.

(viii) By considering congruences mod 5, show that there are no algebraic integers in $\mathbb{Q}(\sqrt{10})$ with norm ±2 and ±3. Prove that $4+\sqrt{10}$ is irreducible in $\mathbb{Q}(\sqrt{10})$. Hence verify that $\mathbb{Q}(\sqrt{10})$ does not have unique factorization.

(ix) Use the fact that $\mathbb{Q}(\sqrt{3})$ is Euclidean to determine algebraic integers α, β in $\mathbb{Q}(\sqrt{3})$ such that $(1+2\sqrt{3})\alpha+(5+4\sqrt{3})\beta=1$.

(x) Prove that the primes in $\mathbb{Q}(\sqrt{2})$ are given by the rational primes $p\equiv\pm3 \pmod 8$, the factors π, π' in the expression $p=\pi\pi'$ appertaining to primes $p\equiv\pm1 \pmod 8$, and by $\sqrt{2}$, together with all their associates.

(xi) Show that if π is a Gaussian prime then the numbers $1, 2, \ldots, N(\pi)$ form a complete set of residues $(\text{mod } \pi)$; that is, show that none of the differences is divisible by π, but that for any Gaussian integer α there is a rational integer a with $1 \leq a \leq N(\pi)$, such that π divides $\alpha - a$. Apply this result to establish the analogue of Fermat's theorem quoted at the end of § 6.

8

Diophantine equations

1 The Pell equation

Diophantine analysis has its genesis in the fertile mind of Fermat. He had studied Bachet's edition, published in 1621, of the first six books that then remained of the famous *Arithmetica*; this was a treatise, originally consisting of thirteen books, written by the Greek mathematician Diophantus of Alexandria at about the third century AD. The *Arithmetica* was concerned only with the determination of particular rational or integer solutions of algebraic equations, but it inspired Fermat to initiate researches into the nature of all such solutions, and herewith the modern theories began.

An especially notorious Diophantine equation, in fact the issue of a celebrated challenge from Fermat to the English mathematicians of his time, is the equation

$$x^2 - dy^2 = 1,$$

where d is a positive integer other than a perfect square. It is usually referred to as the Pell equation but the nomenclature, due to Euler, has no historical justification since Pell apparently made no contribution to the topic. Fermat conjectured that there is at least one non-trivial solution in integers x, y, that is, a solution other than $x = \pm 1$, $y = 0$; the conjecture was proved by Lagrange in 1768. In fact we have already established the result in § 3 of Chapter 7; it was assumed there that d is square-free but the argument plainly holds for any d that is not a perfect square. Now there is a unique solution to the Pell equation in which the integers x, y have their smallest positive values; it is called the fundamental solution. Let x', y' be this solution and put $\varepsilon = x' + y'\sqrt{d}$. Then, by the arguments of § 3 of Chapter 7, we see that all solutions are given by $x + y\sqrt{d} = \pm\varepsilon^n$, where $n = 0, \pm 1, \pm 2, \ldots$. In particular, the equation has infinitely many solutions.

More insight into the character of the solutions is provided by the continued fraction algorithm. First we observe that any solution in positive integers x, y satisfies $x - y\sqrt{d} = 1/(x + y\sqrt{d})$, whence $x > y\sqrt{d}$ and $x - y\sqrt{d} < 1/(2y\sqrt{d})$. This gives $|\sqrt{d} - x/y| < 1/(2y^2)$, and it follows from § 3 of Chapter 6 that x/y is a convergent to $\sqrt{d}$. Now it was noted in § 4 of Chapter 6 that the continued fraction for $\sqrt{d}$ has the form

$$[a_0, \overline{a_1, \ldots, a_m}];$$

the number m of repeated partial quotients is called the period of $\sqrt{d}$. Let p_n/q_n $(n = 1, 2, \ldots)$ be the convergents to $\sqrt{d}$ and let θ_n $(n = 1, 2, \ldots)$ be the complete quotients. We have $x = p_n$, $y = q_n$ for some n, that is $p_n^2 - dq_n^2 = 1$. Here n must be odd for, by § 3 of Chapter 6,

$$\sqrt{d} = \frac{p_n\theta_{n+1} + p_{n-1}}{q_n\theta_{n+1} + q_{n-1}},$$

whence, on recalling that $p_{n-1}q_n - p_nq_{n-1} = (-1)^n$, we obtain

$$q_n\sqrt{d} - p_n = (-1)^n/(q_n\theta_{n+1} + q_{n-1}),$$

and so, for even n, $q_n\sqrt{d} > p_n$. In fact n must have the form $lm - 1$, where $l = 1, 2, 3, \ldots$ when m is even and $l = 2, 4, 6, \ldots$ when m is odd. For the expression for $\sqrt{d}$ above gives

$$(p_n - q_n\sqrt{d})\theta_{n+1} = q_{n-1}\sqrt{d} - p_{n-1},$$

and thus

$$\begin{aligned}(p_n^2 - dq_n^2)\theta_{n+1} &= (q_{n-1}\sqrt{d} - p_{n-1})(q_n\sqrt{d} + p_n)\\ &= (-1)^{n-1}\sqrt{d} + c,\end{aligned}$$

where c is an integer. But $p_n^2 - dq_n^2 = 1$ and n is odd; hence $\theta_{n+1} = \sqrt{d} + c$. Now $\sqrt{d} = a_0 + 1/\theta_1$, where θ_1 is purely periodic, and we have $\theta_{n+1} = a_{n+1} + 1/\theta_{n+2}$. Since $\theta_1 > 1$, $\theta_{n+2} > 1$, we obtain $a_{n+1} = a_0 + c$ and $\theta_1 = \theta_{n+2}$; it follows that $n + 1$ is divisible by m and so n has the form $lm - 1$, as asserted.

We have therefore shown that the only possible positive solutions x, y to the Pell equation are given by $x = p_n$, $y = q_n$, where p_n/q_n is a convergent to $\sqrt{d}$ with n of the form $lm - 1$ as above.

In fact all of these p_n, q_n satisfy the equation and thus they comprise the full set of positive solutions. For, in view of the periodicity of $\sqrt{d}$ we have $\theta_1 = \theta_{n+2}$ for all $n = lm - 1$ as above, and hence

$$\sqrt{d} = \frac{p_{n+1}\theta_1 + p_n}{q_{n+1}\theta_1 + q_n}.$$

But $\sqrt{d} = a_0 + 1/\theta_1$, and on substituting for θ_1 and using the fact that $\sqrt{d}$ is irrational, we obtain

$$p_n = q_{n+1} - a_0 q_n, \qquad p_{n+1} - a_0 p_n = q_n d.$$

On eliminating a_0 we see that

$$p_n^2 - dq_n^2 = p_n q_{n+1} - p_{n+1} q_n,$$

and, since n is odd, it follows that $p_n^2 - dq_n^2 = 1$, as required.

A similar analysis applies to the equation $x^2 - dy^2 = -1$. In this case there is no solution when the period m of $\sqrt{d}$ is even. When m is odd, all positive solutions are given by $x = p_n$, $y = q_n$, where p_n/q_n is a convergent to $\sqrt{d}$ and $n = lm - 1$ with $l = 1, 3, 5, \ldots$. Further, when the equation is soluble, there is a solution in positive integers x', y' of smallest value, known as the fundamental solution, and on writing $\eta = x' + y'\sqrt{d}$, one deduces that all solutions are given by $x + y\sqrt{d} = \pm\eta^n$, where $n = \pm 1, \pm 3, \pm 5, \ldots$; the result is in fact easily obtained on noting that the fundamental solution to $x^2 - dy^2 = 1$ is given by $\varepsilon = \eta^2$. An analogous result holds for the more general equation $x^2 - dy^2 = k$, where k is a non-zero integer. Here, when the equation is soluble, one can specify a finite set of solutions x', y' such that, on writing $\zeta = x' + y'\sqrt{d}$, all solutions are given by $x + y\sqrt{d} = \pm\zeta\varepsilon^n$ with $n = 0, \pm 1, \pm 2, \ldots$.

As an example, consider the equation $x^2 - 97y^2 = -1$. The continued fraction for $\sqrt{97}$ is

$$[9, \overline{1, 5, 1, 1, 1, 1, 1, 1, 5, 1, 18}].$$

Thus the period m of $\sqrt{97}$ is 11 and, since m is odd, the equation is soluble. Indeed the fundamental solution is given by $x = p_{10}$, $y = q_{10}$, where p_n/q_n ($n = 1, 2, \ldots$) denote the convergents to $\sqrt{97}$. Now the first ten convergents to $\sqrt{97}$ are 10, $\frac{59}{6}$, $\frac{69}{7}$, $\frac{128}{13}$, $\frac{197}{20}$, $\frac{325}{33}$, $\frac{522}{53}$, $\frac{847}{86}$, $\frac{4757}{483}$ and $\frac{5604}{569}$. Hence the fundamental solution to $x^2 - 97y^2 = -1$ is $x = 5604$, $y = 569$. Further, if we write $\eta =$

$5604+569\sqrt{97}$ then $\varepsilon=\eta^2$ gives the fundamental solution to $x^2-97y^2=1$; the solution is in fact $x=62\,809\,633$, $y=6\,377\,352$. Incidentally, the continued fraction for $\sqrt{d}$ always has the form

$$[a_0, \overline{a_1, a_2, a_3, \ldots, a_3, a_2, a_1, 2a_0}],$$

as for $\sqrt{97}$ above, and moreover the period m of $\sqrt{d}$ is always odd when d is a prime $p\equiv 1 \pmod 4$. In fact, for such p, the equation $x^2-py^2=-1$ is always soluble. For if x', y' is the fundamental solution to $x^2-py^2=1$ then x' is odd and so $(x'+1, x'-1)=2$; this gives either $x'+1=2u^2$, $x'-1=2pv^2$ or $x'-1=2u^2$, $x'+1=2pv^2$ for some positive integers u, v with $y'=2uv$, whence $u^2-pv^2=\pm 1$, and here the minus sign must hold since $v<y'$.

2 The Thue equation

A multitude of special techniques have been devised through the centuries for solving particular Diophantine equations. The scholarly treatise by Dickson on the history of the theory of numbers (see § 6) contains numerous references to early works in the field. Most of these were of an *ad-hoc* nature, the arguments involved being specifically related to the example under consideration, and there was little evidence of a coherent theory. In 1900, as the tenth of his famous list of 23 problems, Hilbert asked for a universal algorithm for deciding whether or not an equation of the form $f(x_1, \ldots, x_n)=0$, where f denotes a polynomial with integer coefficients, is soluble in integers $x_1, \ldots, x_n$. The problem was resolved in the negative by Matiyasevich, developing ideas of Davis, Robinson and Putnam on recursively enumerable sets. The proof has subsequently been refined to show that an algorithm of the kind sought by Hilbert does not exist even if one limits attention to polynomials in just nine variables, and it seems to me quite likely that it does not in fact exist for polynomials in only three variables. For polynomials in two variables, however, the situation would appear to be quite different.

In 1909, a new technique based on Diophantine approximation was introduced by the Norwegian mathematician Axel Thue. He considered the equation $F(x, y)=m$, where F denotes an irreducible binary form with integer coefficients and degree at least 3, and m is any integer. The equation can be expressed as

$$a_0x^n + a_1x^{n-1}y + \cdots + a_ny^n = m,$$

and this can be written in the form

$$a_0(x - \alpha_1 y) \cdots (x - \alpha_n y) = m,$$

where $\alpha_1, \ldots, \alpha_n$ signify a complete set of conjugate algebraic numbers. Thus if the equation is soluble in positive integers x, y then the nearest of the numbers $\alpha_1, \ldots, \alpha_n$ to x/y, say α, satisfies $|x - \alpha y| \ll 1$. Here we are using Vinogradov's notation; by $a \ll b$ we mean $a < bc$ for some constant c, that is, in this case, a number independent of x and y, and similarly by $a \gg b$ we shall mean $b < ac$ for some such c. Now, for y sufficiently large and for $\alpha \neq \alpha_j$, we have

$$|x - \alpha_j y| = |(x - \alpha y) + (\alpha - \alpha_j)y| \gg y;$$

this gives $|x - \alpha y| \ll 1/y^{n-1}$ whence $|\alpha - x/y| \ll 1/y^n$. But by Thue's improvement on Liouville's theorem mentioned in § 5 of Chapter 6, we have $|\alpha - x/y| \gg 1/y^\kappa$ for any $\kappa > \frac{1}{2}n + 1$. It follows that y is bounded above and so there are only finitely many possibilities for x and y. The argument obviously extends to integers x, y of arbitrary sign, and hence we obtain the remarkable result that the Thue equation has only finitely many solutions in integers. Plainly the condition $n \geq 3$ is necessary here, for, as we have shown, the Pell equation has infinitely many solutions.

The demonstration of Thue just described has a major limitation. Although it yields an estimate for the number of solutions of $F(x, y) = m$, it does not enable one to furnish the complete list of solutions in a given instance or indeed to determine whether or not the equation is soluble. This is a consequence of the ineffective nature of the original Thue inequality on which the proof depends. Some effective cases of the inequality have been derived and, in these instances, one can easily solve the related Thue equation even for quite large values of m; for example, from the result on $\sqrt[3]{2}$ mentioned in § 5 of Chapter 6 we obtain the bound $(10^6|m|)^{23}$ for all solutions of $x^3 - 2y^3 = m$. But still the basic limitation of Thue's argument remains.

Another approach was initiated by Delaunay and Nagell in the 1920s. It involved factorization in algebraic number fields and it enabled certain equations of Thue type with small degree to be solved completely. In particular, the method applied to the equation $x^3 - dy^3 = 1$, where d is a cube-free integer, and it yielded the result that there is at most one solution in non-zero integers x, y. The method was developed by Skolem using analysis in the p-adic domain, and he furnished thereby a new proof of Thue's theorem in the case when not all of the zeros of $F(x, 1)$ are real. The work depended on the compactness property of the p-adic integers and so was generally ineffective, but Ljunggren succeeded in applying the technique to deal with several striking examples. For instance he showed that the only integer solutions (x, y) of the equation $x^3 - 3xy^2 - y^3 = 1$ are $(1, 0)$, $(0, -1)$, $(-1, 1)$, $(1, -3)$, $(-3, 2)$ and $(2, 1)$.

An entirely different demonstration of Thue's theorem was given by Baker in 1968. It involved the theory of linear forms in logarithms (see § 6 of Chapter 6) and it led to explicit bounds for the sizes of all the integer solutions x, y of $F(x, y) = m$; in fact the method yielded bounds of the form $c|m|^{c'}$, where c, c' are numbers depending only on F. Thus, in principle, the complete list of solutions can be determined in any particular instance by a finite amount of computation. In practice the bounds that arise in Baker's method are large, typically of order $10^{10^{500}}$, but it has been shown that they can usually be reduced to manageable figures by simple observations from Diophantine approximation.

3 The Mordell equation

Some profound results relating to the equation $y^2 = x^3 + k$, where k is a non-zero integer, were discovered by Mordell in 1922, and the equation continued to be one of Mordell's major interests throughout his life. The theorems that he initiated divide naturally according as one is dealing with integer solutions x, y or rational solutions. Let us begin with a few words about the latter.

The equation $y^2 = x^3 + k$ represents an elliptic curve in the real projective plane. By a rational point on the curve we shall mean either a pair (x, y) of rational numbers satisfying the equation, or the point at infinity on the curve; in other words, the rational points are given in homogeneous co-ordinates by (x, y, z), where $\lambda x, \lambda y, \lambda z$ are rational for some λ. It had been noted, at least by the time of Bachet, that the chord joining any two rational points on the curve intersects the curve again at a rational point, and similarly that the tangent at a rational point intersects again at a rational point. Thus, Fermat remarked, if there is a rational point on the curve other than the point at infinity, then, by taking chords and tangents, one would expect, in general, to obtain an infinity of rational points; a precise result of this kind was established by Fueter in 1930. It was also well known that the set of all rational points on the curve form a group under the chord and tangent process (see Fig. 8.1); the result is in fact an immediate consequence of the addition formulae for the Weierstrass functions $x=\wp(u)$, $y=\frac{1}{2}\wp'(u)$ that parameterize the curve. Indeed, with this notation, the group law becomes simply the addition of parameters. Mordell proved that the group has a finite basis, that is, there is a finite set of parameters $u_1, \ldots, u_r$ such that all rational points on the curve are given by $u = m_1u_1+\cdots+m_ru_r$, where $m_1, \ldots, m_r$ run through all rational integers. This is equivalent to the assertion that there is a finite set of rational points on the curve such that, on starting from the set and taking all possible chords and tangents, one obtains the totality of rational points on the curve. The demonstration involved an ingenious technique, usually attributed to Fermat, known as the method of infinite descent; we shall refer to the method again in § 4. The work applied more generally to the equation $y^2=x^3+ax+b$ with a, b rational, and so, by birational transformation, to any curve of genus 1. Weil extended the theory to curves of higher genus and the subject subsequently gained great notoriety and stimulated much further research. The latter has been directed especially to the problem of determining the basis elements $u_1, \ldots, u_r$ or at least the precise value of r; this is usually referred to as the rank of the Mordell–Weil group. There is no general algorithm for determining these quantities but they can normally be found in practice. Thus, for instance, Billing proved in 1937 that all the rational points on the curve

$y^2 = x^3 - 2$ are given by mu_1, where u_1 is the parameter corresponding to the point (3, 5) and m runs through all the integers. Since no multiple of u_1 is a period of the associated Weierstrass function, it follows that the equation $y^2 = x^3 - 2$ has infinitely many rational solutions. On the other hand, it is known, for instance, that the equation $y^2 = x^3 + 1$ has only the rational solutions given by $(0, \pm 1)$, $(-1, 0)$ and $(2, \pm 3)$, and that the equation $y^2 = x^3 - 5$ has no rational solutions whatever.

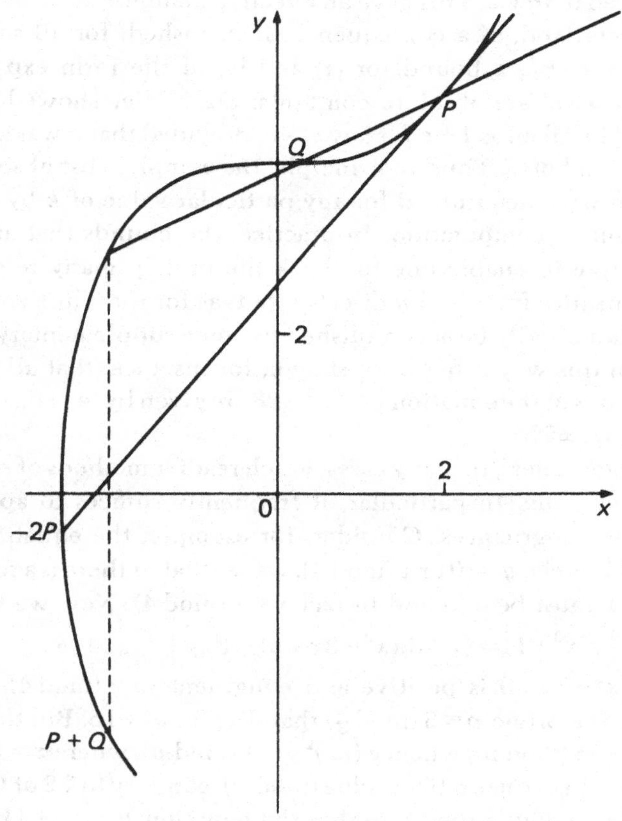

Fig. 8.1. Illustration of the group law on $y^2 = x^3 + 17$. The points P, Q and $P + Q$ are (2, 5), $(\frac{1}{4}, \frac{33}{8})$ and (−2, −3) respectively. The tangent at P meets the curve again at $-2P = (-\frac{64}{25}, -\frac{59}{125})$.

We turn now to integer solutions of $y^2=x^3+k$. Although, initially, Mordell believed that, for certain values of k, the equation would have infinitely many solutions in integers x, y, he later showed that, in fact, for all k, there are only finitely many such solutions. The proof involved the theory of reduction of binary cubic forms and depended ultimately on Thue's theorem on the equation $F(x, y)=m$. Thus the argument did not enable the full list of solutions to be determined in any particular instance. However, the situation was changed by Baker's work referred to in § 2. This gave an effective demonstration of Thue's theorem, and, as a consequence, it furnished, for all solutions of $y^2=x^3+k$, a bound for $|x|$ and $|y|$ of the form $\exp(c|k|^{c'})$, where c, c' are absolute constants; Stark later showed that c' could be taken as $1+\varepsilon$ for any $\varepsilon>0$, provided that c was allowed to depend on ε. Thus, in principle, the complete list of solutions can now be determined for any particular value of k by a finite amount of computation. In practice, the bounds that arise are too large to enable one to check the finitely many remaining possibilities for x and y directly; but, as for the Thue equation, this can usually be accomplished by some supplementary analysis. In this way it has been shown, for instance, that all integer solutions of the equation $y^2=x^3-28$ are given by $(4, \pm 6)$, $(8, \pm 22)$ and $(37, \pm 225)$.

Nevertheless, in many cases, much readier methods of solution are available. In particular, it frequently suffices to appeal to simple congruences. Consider, for example, the equation $y^2=x^3+11$. Since $y^2\equiv 0$ or $1 \pmod 4$, we see that, if there is a solution, then x must be odd and in fact $x\equiv 1 \pmod 4$. Now we have

$$x^3+11=(x+3)(x^2-3x+9)-16,$$

and x^2-3x+9 is positive and congruent to $3 \pmod 4$. Hence there is a prime $p\equiv 3 \pmod 4$ that divides y^2+16. But this gives $y^2\equiv -16 \pmod p$, whence $(yz)^2\equiv -1 \pmod p$, where $z=\frac{1}{4}(p+1)$. Thus -1 is a quadratic residue $\pmod p$, contrary to § 2 of Chapter 4. We conclude therefore that the equation $y^2=x^3+11$ has no solution in integers x, y. Several more examples of this kind are given in Mordell's book (see § 6).

Another typical method of solution is by factorization in quadratic fields. Consider the equation $y^2=x^3-11$. Since $y^2\equiv 0$,

1 or 4 (mod 8) we see that, if there is a solution, then x must be odd. We shall use the result established in § 5 of Chapter 7, that the field $\mathbb{Q}(\sqrt{(-11)})$ is Euclidean and so has unique factorization. We have

$$(y+\sqrt{(-11)})(y-\sqrt{(-11)})=x^3,$$

and the factors on the left are relatively prime; for any common divisor would divide $2\sqrt{(-11)}$, contrary to the fact that neither 2 nor 11 divides x. Thus, on recalling that the units in $\mathbb{Q}(\sqrt{(-11)})$ are ± 1, we obtain $y+\sqrt{(-11)}=\pm\omega^3$ and $x=N(\omega)$ for some algebraic integer ω in the field. Actually we can omit the minus sign since -1 can be incorporated in the cube. Now, since $-11\equiv 1$ (mod 4), we have $\omega=a+\frac{1}{2}b(1+\sqrt{(-11)})$ for some rational integers a, b. Hence, on equating coefficients of $\sqrt{(-11)}$, we see that

$$1=3(a+\tfrac{1}{2}b)^2(\tfrac{1}{2}b)-11(\tfrac{1}{2}b)^3,$$

that is $(3a^2+3ab-2b^2)b=2$. This gives $b=\pm 1$ or ± 2, and so the solutions (a, b) are $(0, -1)$, $(1, -1)$, $(1, 2)$ and $(-3, 2)$. But we have $x=a^2+ab+3b^2$. Thus we conclude that the integer solutions of the equation $y^2=x^3-11$ are $(3, \pm 4)$ and $(15, \pm 58)$. A similar analysis can be carried out for the equation $y^2=x^3+k$ whenever $\mathbb{Q}(\sqrt{k})$ has unique factorization and $k\equiv 2, 3, 5, 6$ or 7 (mod 8).

Soon after establishing his theorem on the finiteness of the number of solutions of $y^2=x^3+k$, Mordell extended the result to the equation

$$y^2=ax^3+bx^2+cx+d,$$

where the cubic on the right has distinct zeros; the work again rested ultimately on Thue's theorem but utilized the reduction of quartic forms rather than cubic. In a letter to Mordell, an extract from which was published in 1926 under the pseudonym X, Siegel described an alternative argument that applied more generally to the hyperelliptic equation $y^2=f(x)$, where f denotes a polynomial with integer coefficients and with at least three simple zeros; indeed it applied to the superelliptic equation $y^m=f(x)$, where m is any integer ≥ 2. The theory was still further extended by Siegel in 1929; in a major work combining his refinement of Thue's inequality, referred to in § 5 of Chapter 6, together with the Mordell–Weil theorem, Siegel succeeded in giving a simple condition for the equation $f(x, y)=0$, where f

is any polynomial with integer coefficients, to possess only finitely many solutions in integers x, y. In particular he showed that it suffices if the curve represented by the equation has genus at least 1. The result was employed by Schinzel, in conjunction with an old method of Runge concerning algebraic functions, to furnish a striking extension of Thue's theorem; this asserts that the equation $F(x, y) = G(x, y)$ has only finitely many solutions in integers x, y, where F is a binary form as in § 2, and G is any polynomial with degree less than that of F.

4 The Fermat equation

In the margin of his well-worn copy of Bachet's edition of the works of Diophantus, Fermat wrote 'It is impossible to write a cube as the sum of two cubes, a fourth power as the sum of two fourth powers, and, in general, any power beyond the second as the sum of two similar powers. For this I have discovered a truly wonderful proof but the margin is too small to contain it'. As is well known, despite the efforts of numerous mathematicians over several centuries, no one has yet succeeded in establishing Fermat's conjecture and there is now considerable doubt as to whether Fermat really had a proof.

Many special cases of Fermat's conjecture have been established, mainly as a consequence of the work of Kummer in the last century. Indeed, as mentioned in Chapter 7, it was Kummer's remarkable researches that led to the foundation of the theory of algebraic numbers. Kummer showed in fact that the Fermat problem is closely related to questions concerning cyclotomic fields. The latter arise by writing the Fermat equation $x^n + y^n = z^n$ in the form

$$(x+y)(x+\zeta y)\cdots(x+\zeta^{n-1}y) = z^n,$$

where ζ is a root of unity. As we shall see in a moment, the case $n = 4$ can be readily treated; thus it suffices to prove that the equation has no solution in positive integers x, y, z when n is an odd prime p. The factors on the left are algebraic integers in the cyclotomic field $\mathbb{Q}(\zeta)$ and, when $p \leq 19$, the field has unique factorization; it is then relatively easy to establish the result. Kummer derived various more general criteria. In particular, he introduced the concept of a regular prime p and proved that Fermat's conjecture holds for all such p; a prime is said to be regular if it does not divide any of the numerators of the first $\frac{1}{2}(p-3)$ Bernoulli numbers, that is, the coefficients B_j in the equation

$$t/(e^t-1)=1-\tfrac{1}{2}t+\sum_{j=1}^{\infty}(-1)^{j-1}B_jt^{2j}/(2j)!.$$

Kummer also established the result for certain classes of irregular primes. Thus, in particular, he covered all $p<100$; there are only three irregular primes in the range and these he was able to deal with separately. The best result to date arising from this approach was obtained by Wagstaff in 1978; by extensive computations he succeeded in establishing Fermat's conjecture for all $p<125\,000$.

Before the work of Kummer, Fermat's equation had already been solved for several small values of n. The special case $x^2+y^2=z^2$ dates back to the Greeks and the solutions (x, y, z) in positive integers are called Pythagorean triples. It suffices to determine all such triples with x, y, z relatively prime and with y even; for if x and y were both odd, we would have $z^2\equiv 2 \pmod 4$ which is impossible. On writing the equation in the form $(z+x)(z-x)=y^2$, and noting that $(z+x, z-x)=2$, we obtain $z+x=2a^2$, $z-x=2b^2$ and $y=2ab$ for some positive integers a, b with $(a, b)=1$. This gives

$$x=a^2-b^2, \qquad y=2ab, \qquad z=a^2+b^2.$$

Moreover, since z is odd, we see that a and b have opposite parity. Conversely, it is readily verified that if a, b are positive integers with $(a, b)=1$ and of opposite parity then x, y, z above furnish a Pythagorean triple with $(x, y, z)=1$ and with y even. Thus we have found the most general solution of $x^2+y^2=z^2$. The first four Pythagorean triples, that is, with smallest values of z, are (3, 4, 5), (5, 12, 13), (15, 8, 17) and (7, 24, 25).

The next simplest case of Fermat's equation is $x^4+y^4=z^4$. This was solved by Fermat himself, using the method of infinite descent. He considered in fact the equation $x^4+y^4=z^2$. If there is a solution in positive integers, then it can be assumed that x, y, z are relatively prime and that y is even. Now (x^2, y^2, z) is a

Pythagorean triple and there exist integers a, b as above such that $x^2=a^2-b^2$, $y^2=2ab$ and $z=a^2+b^2$. Further, b must be even, for otherwise we would have a even and b odd, and so $x^2\equiv -1 \pmod 4$, which is impossible. Furthermore, (x, b, a) is a Pythagorean triple. Hence we obtain $x=c^2-d^2$, $b=2cd$ and $a=c^2+d^2$ for some positive integers c, d with $(c, d)=1$. This gives $y^2=2ab=4cd(c^2+d^2)$. But c, d and c^2+d^2 are coprime in pairs, whence $c=e^2$, $d=f^2$ and $c^2+d^2=g^2$ for some positive integers e, f, g. Thus we have $e^4+f^4=g^2$ and $g\le g^2=a\le a^2<z$. It follows, on supposing that z is chosen minimally at the outset, that the equation $x^4+y^4=z^2$ has no solution in positive integers x, y, z.

The first apparent proof that the equation $x^3+y^3=z^3$ has no non-trivial solution was published by Euler in 1770, but the argument depended on properties of integers of the form a^2+3b^2 and there has long been some doubt as to its complete validity. An uncontroversial demonstration was given later by Gauss using properties of the quadratic field $\mathbf{Q}(\surd(-3))$. The proof is another illustration of the method of infinite descent. By considering congruences $(\text{mod } \lambda^4)$, where λ is the prime $\frac{1}{2}(3-\surd(-3))$ in $\mathbf{Q}(\surd(-3))$, it is readily verified that if the equation $x^3+y^3=z^3$ has a solution in positive integers, then one at least of x, y, z is divisible by λ. Hence, for some integer $n\ge 2$, the equation $\alpha^3+\beta^3+\eta\lambda^{3n}\gamma^3=0$ has a solution with η a unit in $\mathbf{Q}(\surd(-3))$ and with α, β, γ non-zero algebraic integers in the field. It is now easily deduced, by factorizing $\alpha^3+\beta^3$, that the same equation, with n replaced by $n-1$, has a solution as above, and the desired result follows. The equation $x^5+y^5=z^5$ was solved by Legendre and Dirichlet about 1825, and the equation $x^7+y^7=z^7$ by Lamé in 1839. By then, however, the *ad-hoc* arguments were becoming quite complicated and it was not until the fundamental work of Kummer that Fermat's conjecture was established for equations with higher prime exponents.

Numerous results have been obtained concerning special classes of solutions. For instance, Sophie Germain proved in 1823 that if p is an odd prime such that $2p+1$ is also a prime then the 'first case' of Fermat's conjecture holds for p, that is, the equation $x^p+y^p=z^p$ has no solution in positive integers with xyz not divisible by p. Further, Wieferich proved in 1909 that the same conclusion is valid for any p that does not satisfy the congruence $2^{p-1} \equiv 1 \pmod{p^2}$. These results were greeted with great admiration at the times of their discovery. The latter condition is not in fact very stringent; there are only two primes up to $3 \cdot 10^9$ that satisfy the congruence, namely 1093 and 3511. In another direction, an important advance has recently been made by Faltings using algebraic geometry; he has confirmed a long-standing conjecture of Mordell that furnishes, as a special case, the striking theorem that, for any given $n \geq 4$, there are only finitely many solutions to the Fermat equation in relatively prime integers x, y, z. Nonetheless, Fermat's conjecture remains open, and the Wolfskehl Prize, offered by the Academy of Sciences in Göttingen in 1908 for the first demonstration, still awaits to be conferred. The prize originally amounted to 100 000 DM, but, alas, the sum has been much eroded by inflation.

5 The Catalan equation

In 1844, Catalan conjectured that the only solution of the equation $x^p - y^q = 1$ in integers x, y, p, q, all >1, is given by $3^2 - 2^3 = 1$. The conjecture has not yet been established but a notable advance towards a demonstration was made by Tijdeman in 1976. He proved, by means of the theory of linear forms in logarithms (see § 6 of Chapter 6), that the equation has only finitely many integer solutions and all of these can be effectively bounded. Thus, in principle, Tijdeman's work reduces the problem simply to the checking of finitely many cases; however, at present, the bounds furnished by the theory are too large to make the computation practical.

To illustrate the approach, let us consider the simpler equation $ax^n - by^n = c$, where a, b and $c \neq 0$ are given integers, and we seek to bound all the solutions in positive integers x, y and $n \geq 3$. We can assume, without loss of generality, that a and b are positive, and it will suffice to treat the case $y \geq x$. The equation can be written in the form $e^\Lambda - 1 = c/(by^n)$, where

$$\Lambda = \log(a/b) + n \log(x/y).$$

Now we can suppose that $y^n > (2c)^2$, for otherwise the solutions can obviously be bounded; then we have $|e^{\Lambda}-1|<1/(2y^{\frac{1}{2}n})$. But, for any real number u, the inequality $|e^u-1|<\frac{1}{2}$ implies that $|u|\leq 2|e^u-1|$. Hence we obtain $|\Lambda|<y^{-\frac{1}{2}n}$, that is $\log|\Lambda|<-\frac{1}{2}n\log y$. On the other hand, by the theory of linear forms in logarithms, we have $\log|\Lambda|\gg -\log n\log y$, where the implied constant depends only on a and b. Thus we see that $\frac{1}{2}n\ll\log n$, whence n is bounded in terms of a and b. The required bounds for x and y now follow from the effective result of Baker on the Thue equation referred to at the end of § 2.

The work of Tijdeman on the equation $x^p-y^q=1$ runs on the same lines. One can assume that p, q are odd primes and then, by elementary factorization, one obtains $x=kX^q+1$, $y=lY^p-1$ for some integers X, Y, where k is 1 or $1/p$ and l is 1 or $1/q$. Plainly we have $|p\log x-q\log y|\ll y^{-q}$, and substituting for x and y on the left yields a linear form

$$\Lambda=p\log k-q\log l+pq\log(X/Y)$$

for which $|\Lambda|$ is small; similar forms arise by substituting for just one of x and y. The theory of linear forms in logarithms now furnishes the desired bounds for p and q, and those for x and y then follow from an effective version of the result on the superelliptic equation referred to in § 3.

Several instances of Catalan's equation were solved long before the advent of transcendence theory. Indeed, in the Middle Ages, Leo Hebraeus had already dealt with the case $x=3$, $y=2$ and, in 1738, Euler had solved the case $p=2$, $q=3$. The case $q=2$ was treated by V.A. Lebesgue in 1850, the cases $p=3$ and $q=3$ by Nagell in 1921, the case $p=4$ by S. Selberg in 1932 and the case $p=2$, which includes the result for $p=4$, by Chao Ko in 1967. Moreover Cassels proved in 1960 that if p, q are primes, as one can assume, then p divides y and q divides x. Let us convey a little of the flavour of these works by proving that the equation $x^5-y^2=1$ has no solution in integers except $x=1$, $y=0$. We shall use the unique factorization property of the Gaussian field. Clearly, since $y^2\equiv 0$ or $1\pmod 4$, we have x odd and y even. The equation can be written in the form $x^5=(1+iy)(1-iy)$ and, since x is odd, the factors on the right are relatively prime. Thus we have $1+iy=\varepsilon\omega^5$, where ε is a Gaussian unit and ω is a Gaussian integer. Now $\varepsilon = \pm 1$ or $\pm i$, and so $\varepsilon = \varepsilon^5$. Hence, on writing $\varepsilon\omega = u + iv$, where u, v are rational integers, we have $1 + iy = (u + iv)^5$. This gives $u^5 - 10u^3v^2 + 5uv^4 = 1$, whence $u = \pm 1$ and $1 - 10v^2 + 5v^4 = \pm 1$. It follows easily that $u = 1$, $v = 0$ and so $x = 1$, $y = 0$, as required. The argument can readily be extended to establish Lebesgue's more general theorem that, for any odd prime p, the equation $x^p - y^2 = 1$ has no solution in positive integers.

A particularly striking result on an exponential Diophantine equation rather like those of Fermat and Catalan was obtained by Erdős and Selfridge in 1975; they proved that a product of consecutive integers cannot be a perfect power, that is, the equation

$$y^n = x(x+1)\cdots(x+m-1)$$

has no solution in integers x, y, m, n, all >1.

6 Further reading

As remarked in § 2, an excellent source for early results on Diophantine equations is Dickson's *History of the theory of numbers* (Washington, 1920; reprinted Chelsea Publ. Co., New York). Another good reference work is Skolem's *Diophantische Gleichungen* (Springer-Verlag, Berlin, 1938).

For general reading, Mordell's *Diophantine equations* (Academic Press, London and New York, 1969) is to be highly recommended; the author was one of the great contributors to the subject and, as one would expect, he covers a broad range of material with clarity and considerable skill.

As regards special topics, Ribenboim's *13 Lectures on Fermat's last theorem* (Springer-Verlag, Berlin, 1979) is a very fine monograph. There is also a good discussion of the Fermat equation in Borevich and Shafarevich *Number theory* (Academic Press, New York and London, 1966). Several other books on number theory contain valuable sections on Diophantine equations; this applies especially to Nagell's *Introduction to number theory* (Wiley, New York, 1951). For an account of recent results on the effective solution of Diophantine equations see Baker's *Transcendental number theory* (Cambridge U.P., 2nd edn, 1979).

The theorem of Schinzel referred to at the end of § 3 appeared in *Comment. Pontificia Acad. Sci.* **2** (1969), no. 20, 1–9. The theorem of Faltings referred to at the end of § 4 appeared in *Invent. Math.* **73** (1983), 349–66. The theorem of Erdős and Selfridge referred to at the end of § 5 appeared in the *Illinois J. Math.* **19** (1975), 292–301.

7 Exercises

(i) Prove that, if (x_n, y_n), with $n = 1, 2, \ldots$, is the sequence of positive solutions of the Pell equation $x^2 - dy^2 = 1$, written according to increasing values of x or y, then x_n and y_n satisfy a recurrence relation $u_{n+2} - 2au_{n+1} + u_n = 0$, where a is a positive integer. Find a when $d = 7$.

(ii) Determine whether or not the equation $x^2 - 31y^2 = -1$ is soluble in integers x, y.

(iii) Show that if p, q are primes $\equiv 3 \pmod 4$ then at least one of the equations $px^2 - qy^2 = \pm 1$ is soluble in integers x, y.

(iv) Prove, by congruences, that if a, c are integers with $a > 1$, $c > 1$ and $a + c \leq 16$, then the equation $x^4 - ay^4 = c$ has no solution in rationals x, y.

(v) Show that the equation $x^3 + 2y^3 = 7(z^3 + 2w^3)$ has no solution in relatively prime integers x, y, z, w.

(vi) By considering the intersection of the quartic surface $x^4 + y^4 + z^4 = 2$ with the line $y = z - x = 1 - tx$, where t is a parameter, show that the equation $x^4 + y^4 + z^4 = 2w^4$ has infinitely many solutions in relatively prime integers x, y, z, w.

(vii) Solve the equation $y^2 = x^3 - 17$ in integers x, y by considering the factors of $x^3 + 8$.

(viii) Solve the equation $y^2 = x^3 - 2$ in integers x, y by factorization in $\mathbb{Q}(\sqrt{(-2)})$.

(ix) Prove, by the method of infinite descent, that the equation $x^4 - y^4 = z^2$ has no solution in positive integers x, y, z.

(x) Prove that the equation $x^4 - 3y^4 = z^2$ has no solution in positive integers x, y, z. Deduce that the equation $x^4 + y^4 = z^3$ has no such solution with $(x, y) = 1$. (For the first part see Pocklington, *Proc. Cambridge Phil. Soc.* **17** (1914), 108–21.)

(xi) By considering $(x+1)^3 + (x-1)^3$, show that every integer divisible by 6 can be represented as a sum of four integer cubes. Show further that every integer can be represented as a sum of five integer cubes.

(xii) Prove, by factorization in $\mathbb{Q}(\sqrt{(-7)})$, that the equation $x^2 + x + 2 = y^3$ has no solution in integers x, y except $x = 2$, $y = 2$ and $x = -3$, $y = 2$. Verify that the equation $x^2 + 7 = 2^{3k+2}$ has no solution in integers k, x with $k > 1$. (This is a special case of a conjecture of Ramanujan to the effect that the equation $x^2 + 7 = 2^n$ has only the integer solutions given by $n = 3, 4, 5, 7$ and 15. The conjecture was proved by Nagell; for a demonstration see Mordell's book, page 205.)

Index

algebraic integer, 62
algebraic number, 51
algebraic number theory, 61, 71
'almost all' natural numbers, 13
'almost all' real numbers, 48, 53
Apéry, R., 54
Apostol, T. M., 16
arithmetical functions, 8–26
Artin, E., 33, 71
automorph of a binary form, 38
average order of arithmetical functions, 12–14

Bachet, C. G., 39, 74, 80, 84
Bachmann, P., 33
Baker, A., 52, 54, 59, 67, 71, 79, 82, 88, 89
Barnes, E. S., 68
Bertrand's postulate, 5
Billing, G., 80
Blichfeldt's result, 56
Borevich, Z. I., 33, 89
box (pigeon-hole) principle, 43

canonical decomposition, 3
Cantor, G., 53
Cassels, J. W. S., 33, 41, 59, 71, 88
Catalan equation, 87–91
Chandrasekharan, K., 16
Chao Ko, 88
Chatland, H., 68, 71
Chen's theorem, 6, 7
Chevalley's theorem, 34
Chinese remainder theorem, 18, 22, 39
class field theory, 33
class number of quadratic forms, 36
complete quotients of continued fraction, 45
complete set of residues, 18
congruences, 18–26
conjugates of an algebraic number, 61
continued fractions, 3, 44–46, 53, 59, 75
convergents of continued fraction, 45
convex body theorem, 56
coprime numbers, 2
critical strip of zeta-function, 15
cyclotomic field, 62, 84

Davenport, H., 7, 41, 68, 71
Davis, M., 7, 77
Dedekind, R., 62
definite and indefinite forms, 35
Delaunay, B., 79
determinant of a lattice, 57
Dickson, L. E., 68, 77, 89
difference between consecutive primes, 6, 15
Diophantine approximation, 43–60
Diophantine equations, 52, 74–91
Diophantus, 74, 84
Dirichlet, G. L., 86
Dirichlet's theorem on arithmetical progressions, 6
Dirichlet's theorem on Diophantine approximation, 43, 58, 64
Dirichlet's theorem on units, 65
discriminant of number field, 62
discriminant of quadratic form, 35
divisibility, 1–7
division algorithm, 1, 6
Dyson, F. J., 52

equivalence of quadratic forms, 35
Eratosthenes sieve, 6
Erdős, P., 89, 90
Euclid, 4, 6
Euclidean fields, 67–69
Euclid's algorithm 3, 6, 46, 67
Euler, L., 5, 6, 29, 38, 74, 86, 88

Euler product of zeta-function, 15
Euler's constant γ, 13, 54
Euler's criterion, 27
Euler's identity, 53
Euler's theorem on congruences, 19
Euler's (totient) function ϕ, 9

Faltings, G., 87, 90
Fermat, P., 5, 38, 74, 80, 84
Fermat equation, 84–87
Fermat primes, 5
Fermat's last theorem, 54, 61, 62, 89
Fermat's method of infinite descent, 40, 80, 85, 86
Fermat's theorem on congruences, 19, 70
Fibonacci sequence, 60
field of residues mod p, 19
fractional part of a number, 8
Fröhlich, A., 33, 71
Fueter, R., 80
functional equation of zeta-function, 15
fundamental theorem of arithmetic, 3, 6, 62, 66

Gauss, C. F., 29, 32, 40, 67, 86
Gaussian integer, 69
Gaussian field, 63, 69–71, 88
Gaussian prime, 69
Gauss' lemma, 28
Gelfond, A. O., 52
Gelfond–Schneider theorem, 54
Geometry of Numbers, 56, 59
Germain, Sophie, 86
Goldbach's conjecture, 6
greatest common divisor, 2

Hadamard, J., 5
Halberstam, H., 7
Hardy, G. H., 7, 15, 16, 25, 59
Hardy–Littlewood circle method, 6, 40, 41
Hecke, E., 71
Heilbronn, H., 68
Hermite, C., 53
Hilbert, D., 40, 54
Hilbert's seventh problem, 54
Hilbert's tenth problem, 5, 7, 77
Hurwitz's theorem, 46, 51
hyperelliptic equation, 83

indices, 24
Inkeri, K., 68
integral basis of number field, 62
integral part of a number, 8
irreducible elements in number field, 62

Jacobi's symbol, 31

Khintchine, A. Y., 59
Kronecker's theorem, 59
Kummer, E. E., 62, 84, 86

Lagrange, J. L., 20, 39, 48, 74
Lagrange's theorem, 21
Lambert series, 17
Lamé, G., 86
Landau, E., 6, 7, 41, 71
Lang, S., 71
lattice, 29, 57
law of quadratic reciprocity, 29, 33
Lebesgue, V. A., 88, 89
Legendre, A. M., 29, 40, 86
Legendre's symbol, 27
Leo Hebraeus, 88
Lindemann, F., 53, 55
linear congruence, 18
linear forms in logarithms, 53, 54, 79, 88
linear independence, 57, 58
Liouville's theorem, 50–53, 78
Ljunggren, W., 79
lowest common multiple, 4

Markoff chain, 47
Matiyasevich, Y. V., 7, 77
Mersenne prime, 5, 14, 33
Minkowski's conjecture, 59
Minkowski's theorem, 56–59
Möbius function μ, 10
Möbius inversion formulae, 10
Mordell, L. J., 79, 80, 82, 83, 87, 89
Mordell equation, 79–84
Mordell–Weil theorem, 80, 83
multiplicative functions, 9

Nagell, T., 7, 25, 79, 88, 89, 91
Narkiewicz, W., 71
Niven, I., 41, 59
norm in quadratic fields, 62
number of divisors τ, 11

Oppenheimer, H., 68

partial quotients of continued fraction, 45

Peano axioms, 1, 6
Pell equation, 38, 50, 65, 74–77, 78
perfect number, 14
Perron, O., 59, 68
prime, 3
prime-number theorem, 4, 15
primes in quadratic fields, 65
primitive root, 22
principal form, 35
principle of mathematical induction, 1
purely periodic continued fractions, 50
Putnam, H., 7, 77
Pythagorean triples, 85

quadratic congruence, 27
quadratic fields, 61–73
quadratic forms, 35–42
quadratic irrationals, 48
quadratic residues, 27–34
quaternions, 39

Ramanujan, S., 91
Ramanujan's sum, 17
rational approximations, 46–48
Rédei, L., 68
reduced set of residues, 19
reduction of quadratic forms, 36
regular prime, 85
relatively prime numbers, 2
Remak, R., 68
representation by binary forms, 37
residue class, 18
Ribenboim, P., 89
Richert, H. E., 7
Riemann hypothesis, 15, 54
Riemann zeta-function, 5, 14–16, 54
Robinson, J., 7, 77
Roth, K. F., 52, 58
Runge, C., 84

Schinzel, A., 84, 90
Schmidt, W. M., 58, 59
Selberg, S., 88
Selfridge, J. L., 89, 90
Shafarevich, I. R., 33, 89
Siegel, C. L., 52, 58, 83
sieve methods, 6, 7
Skolem, T., 79, 89
squaring the circle, 53
standard factorization, 4
Stark, H. M., 7, 67, 82
Stewart, I., 71
sum of divisors σ, 12
sum of four squares, 39
sum of three squares, 40
sum of two squares, 38, 70
superelliptic equation, 83, 88
Swinnerton-Dyer, H. P. F., 68
Sylvester's argument, 10

Tall, D., 71
Tate, J., 33
Thue, A., 52, 58, 77
Thue equation, 77–79
Tijdeman, R., 87, 88
Titchmarsh, E. C., 16
transcendental numbers, 51, 52, 53–56, 58, 59

unimodular substitution, 35
unique factorization domain, 62, 66
units in number fields, 62, 63–65

Vallée Poussin, C. J. de la, 5
Vaughan, R. C., 41
Vinogradov, I. M., 6, 25, 78
von Mangoldt's function Λ, 17

Wagstaff, S. S., 85
Waring, E., 20
Waring's problem, 40, 41
Weil, A., 33, 71, 80
Wieferich, A., 87
Wilson's theorem, 20
Wolfskehl Prize, 87
Wolstenholme's theorem, 26
Wright, E. M., 7, 16, 25, 59

Zuckerman, H. S., 41, 59